WA 1166316 2

D0421735

Exploring

CALCULUS

**UNIVERSITY OF GLAMORGAN
LEARNING RESOURCES CENTRE**

Pontypridd, Mid Glamorgan, CF37 1DL
Telephone: Pontypridd (01443) 482626

Books are to be returned on or before the last date below

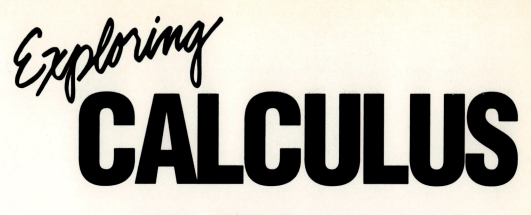

Exploring CALCULUS

with *Mathematica*®

FOR THE MACINTOSH INTERFACE

James K. Finch
Millianne Lehmann
University of San Francisco

▲

ADDISON-WESLEY PUBLISHING COMPANY
Reading, Massachusetts • Menlo Park, California • New York
Don Mills, Ontario • Wokingham, England • Amsterdam • Bonn
Sydney • Singapore • Tokyo • Madrid • San Juan • Milan • Paris

Learning Resources
Centre

Copyright © 1992 by Addison-Wesley Publshing Company, Inc. All rights reserved. No part of this publication may be reproduced, stored in a retrieval system, or transmitted, in any form or by any means, electronic, mechanical, photocopying, recording, or otherwise, without the prior written permission of the publisher.
Printed in the United States of America.

Mathematica® is a registered trademark of Wolfram Research, Inc. Mathematica is not associated with Mathematica, Inc., Mathematica Policy Research, Inc., or MathTech, Inc.

ISBN 0-201-55572-7

5 6 7 8 9 10-CRS-99 98 97 96

Preface

Sophisticated symbolic and numerical manipulation computer programs, such as *Mathematica*, transform the study of calculus. It need no longer be restricted to those problems and applications which can be approached symbolically with a minimum of computational effort. In the old days every algebraic expression factored easily, angles were always multiples of 15 degrees, graphs had no more than three critical points, and numerical methods were seldom treated. Those times are gone. So is the necessity for lectures and homework assignments full of algorithmic drill. *Mathematica* and other similar systems have changed all that. These high-powered programs can integrate, differentiate, and solve terribly complicated algebraic equations. They draw graphs in two and three dimensions, construct tables, calculate limits, and solve differential equations to name only a few of their most important capabilities relevant to beginning calculus.

Pick a problem at random from a standard calculus text and chances are that *Mathematica* will solve it quickly and accurately and with code that in most cases mimics conventional mathematical usage to a remarkable degree. The small price one must pay is the time it takes to master the basic rules of *Mathematica* syntax and of the computer interface. It turns out to be a much less daunting task than one might imagine. Calculus can be handled with a relatively small number of commands; a few dozen are mentioned in this book and only 20 or so of those are used repeatedly. As for the interface, the typical student of calculus unfamiliar with the Apple Macintosh computer masters it in a two or three sessions.

The rewards are enormous. The traditional calculus course has been by and large constrained to a symbolic presentation of the basic ideas of the calculus and the problems calculus is designed to solve. We are not suggesting that symbols be ignored. After all, the great strength of the calculus is that it reduces complicated and sophisticated infinite processes to a brilliantly designed symbolic system. However, the meaning behind the symbolism is another matter. Concepts central to the calculus such as function and limit, continuity and differentiability, instantaneous rates of change, to name a few, are greatly illuminated by graphical and numerical treatments. These treatments help build intuition. They promote understanding by illustrating and reinforcing ideas that for beginners are deep and difficult. The graphical and numerical approach is greatly facilitated by *Mathematica*.

Each of the chapters in this book is called an Exploration. The book contains twenty-six of these *Mathematica*-based explorations of topics from first year calculus. They are addressed to the beginning student who the authors assume is reading the book and using *Mathematica* as part of calculus course work. Each is designed for a two-hour *Mathematica* session during which the reader works alone or in a group of two or three on the material introduced. (An Exploration can also be used as the basis for a classroom demonstration.)

We have included a basic introduction to the Macintosh computer most of which is contained in the first two explorations. We deliberately chose to work directly with *Mathematica* and not try to put a simplified user interface around it. The syntax of *Mathematica* causes little difficulty after an initial adjustment period and the real advantage of this approach is that the student is then able to use the whole power of *Mathematica* in higher level courses and in the many science courses for which calculus is a prerequisite. There is a long-run payoff in avoiding the restriction of a shell designed specifically for calculus.

Exploring Calculus with Mathematica was written in *Mathematica* 2.0 and we assume that the reader is running that version with full menus on a stand-alone Macintosh computer. Five megabytes of RAM are needed for those explorations which contain animations or 3D graphics. Plot options are sometimes set for the inclusion of color in a graph. In these cases alternate gray level specifications are included. All of the graphics will display on a monochrome monitor. Color, of course, produces a more interesting picture. Experienced *Mathematica* users can start anywhere, but the first two chapters are a must for the novice.

This book is accompanied by the *Exploring Calculus* disk, which includes all of the book's input cells organized by Exploration number. It is never necessary to type in an input statement displayed in this book if the reader does not wish to. They can be copied from the disk into a *Mathematica* file called a notebook. In addition to the input statements, the *Exploring Calculus* disk contains several packages which are used in a number of the explorations. For more details see Appendix 3.

Our thanks to the National Science Foundation and the College of Liberal Arts and Sciences, University of San Francisco, for providing financial support for this project. Our appreciation goes also to the many individuals who have given advice and encouragement: from the University–our colleagues Clifton Albergotti, Allan Cruse, Stanley Nel, Robert Wolf, and Natalia Zatov and our teaching assistants and interns Kelley

Absher, Paul Blobner, Hoang Dinh, John Heng, Edward Jenner, Ellen Kohrdahl, Kitty Ng, Marta Nichols, T.J. Willis, and Doug Pager; from Wolfram Research–Brad Horn and Jamie Peterson; and from Addison-Wesley–Beth Perry, Allan Wylde, and Jerome Grant.

James K. Finch
Millianne Lehmann

University of San Francisco
October 1991

Contents

x   Contents

1

Welcome to *Mathematica*!

Getting Started

In this section you will learn how to use *Mathematica* to perform electronically many of the paper and pencil mathematics algorithms which you were taught in precalculus courses. Before you can begin you must turn on your Apple Macintosh computer, and open up a *Mathematica* notebook. If you are a Macintosh novice, you may need help in finding the *Mathematica* folder and, since the procedures for starting *Mathematica* vary widely from system to system, you will also need to find out how to open a notebook on your particular machine.

When you have succeeded in entering *Mathematica* you should see a blank, bordered rectangle on the screen with the label "Untitled–1" written in the middle of its top edge.

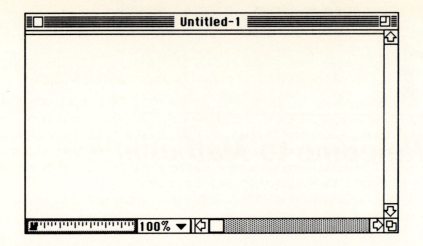

The rectangle is called a *window* in Macintosh-ese and that top border is referred to as its *title bar*. This is your first *Mathematica* file, or *notebook*, as it is usually called. In it you will do calculations, type text, and draw graphs much as you would in an ordinary paper notebook.

THE CURSOR

The mouse, which is attached to your Macintosh, is a pointing device. Move the mouse in various directions while watching the screen, and you will see a small moving image. It is called the *cursor*. It will change in size and shape depending on where you position it on the screen. Try this: Use the mouse to move the cursor from one side of the screen to the other. When it is inside your notebook, it has the shape of a horizontal I-beam.

On the borders of the window and outside the window it changes into an arrow.

The arrow and I-beam are just two of many forms that the cursor can take on. In time, you will become familiar with them all. Meanwhile, the next paragraph tells how you can use the arrow cursor to change the way the window looks on the screen. More about the I-beam later.

CHANGING THE SIZE AND POSITION OF A WINDOW

Your notebook window can be moved up and down and from side to side. Use the mouse to move the cursor to the title bar. The title bar looks like this:

Untitled-1

As the cursor passes into the title bar it turns into an arrow. Position the point of the arrow in the title bar (you are *point*ing to the title bar in the language of Macintosh users). Press and hold the mouse button and drag the mouse in the direction in which you wish to move the window. A dotted outline of the window will appear, and it will move as you move the cursor. Release the mouse button when you have placed the outline where you want the window to be. The window will move to that position.

If you want to change the size of the window, use the mouse to point to the overlapped squares in the window's lower right corner.

Remember, the arrow point must be inside the overlapped squares. Now, press and hold the mouse button and drag the mouse up and to the left to shrink the window, or down and to the right to enlarge it.

Practice repositioning and reshaping the window.

WHAT TO DO IF THE SCREEN SUDDENLY CHANGES TO A MOVING DISPLAY

Your Macintosh may be running what is called a *screen saver*. When a stationary image is left in place too long, its outlines can be permanently burned onto the screen. A screen saver protects against this by automatically producing moving images if no key is pressed for several minutes. The sudden appearance of these images, which can be anything from scrolling lines to flying toasters, can be disconcerting if you are not expecting them. Don't panic. You can return the screen to the normal display by pressing a key or clicking the mouse.

Using *Mathematica* to Do Arithmetic

Give *Mathematica* the problem of adding 5 to 7. Use the mouse to position the I-beam cursor in your notebook, click the mouse button, and type **5+7**. (If you make a mistake typing, use the *delete* key to backspace and erase.) Then press the *enter* key. If there is no *enter* key, hold down the *shift* key and press *return*. The *return* key by itself will not do the trick; when *return* is used alone without the *shift* key it simply moves the cursor to the next line. The *enter* key or *shift-return* tells *Mathematica* that you have finished typing a command and you wish to have it executed.

Below you can see what the computation **5 + 7** looks like on the screen. The phrases **In[1]** and **Out[1]** are written in by *Mathematica*. Do not type them yourself. *Mathematica* will automatically number your commands for you in sequential order as you enter them.

```
In[1]:=
        5+7
```

```
Out[1]=
        12
```

LOADING THE KERNEL

You may have to wait for *Mathematica* to load its *kernel* before you get this result. The kernel, which is the heart of *Mathematica*, is the 300,000 lines of code which permit it to do mathematics. If the sum is returned immediately then the kernel was loaded earlier in the current *Mathematica* session. If the sum is not returned immediately, look up at the top right of the screen. Do you see the word **Starting...**?

 ⚫ File Edit Cell Graph Find Action Style Window Starting...

If you do, the kernel is being loaded. If **Starting...** is not there and *Mathematica* does not return the sum, any delay is due to some other cause. Did you press the *enter* key?

CELLS

When the addition problem is complete, look at the right side of your notebook. You will see three brackets opposite the arithmetic calculation you just did.

The top inside bracket encloses the expression **5 + 7**, and it marks the boundary of the *input cell*. The bottom inside bracket encloses the *output cell*. The input and output cells for the calculation are grouped together by the outer bracket. Cells are the fundamental building blocks of a *Mathematica* notebook. The input and output cells you just saw are only two of the many different types of cells which *Mathematica* can create. Learning about cell types is part of mastering *Mathematica*, and you've made a beginning. More about cells later.

MULTIPLICATION AND DIVISION

Try a multiplication problem next. In *Mathematica* the multiplication sign is * (*shift* 8). Multiplication is also indicated by a space. Type the command **2*3** and press the *enter* key.

```
In[2]:=
        2*3
```

```
Out[2]=
        6
```

Now, type **2**, press the space bar, type **3**, and press the *enter* key.

```
In[3]:=
        2 3
```

```
Out[3]=
        6
```

The division sign is **/**, which is found next to the right-hand *shift* key. The subtraction sign is the hyphen. Exponentiation is indicated by **^** (*shift* 6).

MORE ARITHMETIC EXAMPLES

The input-output cells shown in the next paragraph give some additional examples of arithmetic done in *Mathematica*, and they also illustrate the use of the **N** command with which you can control the form in which numerical output is displayed. For practice, you should type and *enter* each of these input cells and compare your output cell to the corresponding one displayed in the text.

Here is the evaluation of a more complicated arithmetic expression:

```
In[4]:=
        (2 + 4)^2/5
```

```
Out[4]=
        36
        --
        5
```

Notice that *Mathematica* squared 6 first and then divided by 5. The result will be quite different if the fraction 2/5 is enclosed in parentheses.

```
In[5]:=
        (2 + 4)^(2/5)
```

```
Out[5]=
        2/5
        6
```

You see that *Mathematica* left the exponent in fraction form. When an arithmetic problem involves only integers, *Mathematica* will always return exact answers instead of decimal approximations. If you wish the result in decimal form, an easy way to get it is to insert a decimal point in one of the numbers in the problem.

```
In[6]:=
        (2. + 4)^(2/5)
```

```
Out[6]=
        2.04767
```

Another way is to use the **N** command.

```
In[7]:=
        N[(2 + 4)^(2/5)]
```

```
Out[7]=
        2.04767
```

The **N** command can also be employed to control the number of digits shown.

```
In[8]:=
        N[(2 + 4)^(2/5), 15]
```

```
Out[8]=
        2.04767251107922
```

Don't leave out the comma. If you do, *Mathematica* will treat the **15** as a multiplier. Here are the first 30 digits of π:

```
In[9]:=
        N[Pi,30]
```

```
Out[9]=
        3.14159265358979323846264338328
```

The command must be typed *exactly* as it appears above—the **N** and **P** must be capitalized and square brackets [ ] must be placed after the **N** and following the **30**.

The next example is from "higher" arithmetic: Find the prime factors of 1400.

```
In[10]:=
        FactorInteger[1400]
```

```
Out[10]=
        {{2, 3}, {5, 2}, {7, 1}}
```

This output tells you that $1400 = 2^3 5^2 7$.

Debugging and Editing

EDITING CELLS

As you were copying input lines and entering them, you probably made a few mistakes. If not, you eventually will. Everyone does. Frequently the mistake creates what is called an *error in syntax*, meaning that the form of the expression entered does not conform to the rules of construction that *Mathematica* understands. You can tell when this has happened because *Mathematica* will scold you with a beep (if your monitor's sound is turned on) or an error message or both.

Suppose you made a mistake typing in the "π" command above and entered `N(pi,30)` instead of `N[Pi,30]`. Give it a try: Enter `N(pi,30)` and see what happens. *Mathematica* cannot "understand" the command because the syntax is wrong, so it will position the blinking vertical cursor in the command at the point where it became confused.

`N(pi|,30]`

By placing the cursor in the flawed command, *Mathematica* is inviting you to fix it so that it can be executed. As it turns out, this error is easy to fix. With the blinking cursor between the **i** and the comma, press and hold the mouse button and drag across the three characters `(pi` and then type `[Pi`. Press the *enter* key and the command will be executed. You should now see the first 29 decimal places of π.

Sometimes you will make a mistake typing in a command, but *Mathematica* will be able to execute it anyway. Of course, the output will not be what you had in mind at all. For example, suppose you left out the **i** in `N[Pi, 30]` and entered the command `N[P, 30]`.

```
In[11]:=
        N[P,30]
```

```
Out[11]=
         P
```

To get the output you want, you need to insert an **i** following the **P** in the input cell. Use the mouse to move the cursor into the flawed input cell. Notice that the cursor flips to a vertical I-beam once it is inside the cell.

Ɪ

Position the vertical cursor after the **P** and click. The cursor will become a blinking vertical bar.

`N[P|,30]`

Type in the missing **i** and press *enter*.

When *Mathematica* isn't responding the way you think it should and you know you must change the input cell but don't know how, here are some things to consider. Expressions such as **N** and **FactorInteger** are examples of built-in *Mathematica functions* or *commands*. The first letter of such expressions is *always* capitalized. Often other letters in the command, such as the **I** in **FactorInteger**, are also capitalized. *Mathematica* will not recognize a command if it is even slightly misspelled.

A function is always followed by an open square bracket. The expressions typed between the open and close square brackets are called the *arguments of the function*. *Mathematica* depends on square brackets to tell it where the arguments begin and where they end. No other grouping symbol will do. So, if you type (or { by mistake when you mean [, *Mathematica* will beep at you and reposition the cursor in the command so that you can edit it.

Grouping symbols must be closed. If your command contains [, it must also contain].

Make sure that there are no extra characters in the input cell, only those that are part of the command you are trying to execute. Delete anything else in the cell.

CREATING A NEW CELL

Another way to handle the problem of mistyped input cells is to simply ignore the flawed command, create a new cell, and retype the entire command. You can create a new cell by using the mouse to move the cursor down the screen until it becomes a horizontal I-beam. Click at that point on the screen. *Mathematica* will draw a line across the screen. Begin typing and the characters will be contained in a new cell.

CUTTING A CELL

If you want to get rid of a cell, maybe because it is just too full of mistakes, follow this procedure. First point to the bracket of the cell you wish to erase and click the mouse button. The cell bracket will become a filled-in rectangle.

This process is called *selecting a cell*.

Now, look up at the very top left hand-corner of the screen and find the icon. This icon and the words to the right of it constitute *Mathematica*'s main menu bar.

 File   Edit   Cell   Graph   Find   Action   Style   Window

For the purpose of cutting a cell, you are interested only in the word **Edit**. Use the mouse to move the cursor to this word. Notice again that the cursor turns into an arrow as soon as it crosses into the menu bar. Place the point on the word **Edit**. Press and hold the mouse button and the **Edit** menu will drop down the screen.

```
       ≰  File  Edit
                Can't Undo            ⌘z

                Cut                   ⌘H
                Copy                  ⌘c
                Paste                 ⌘v
                Clear                 ⌘H
                Paste and Discard
                Convert Clipboard...  ⌘m

                Select All Cells      ⌘A

                Nesting               ▶

                Preferences           ▶
                ✓Long Menus
```

While keeping the mouse button depressed, drag the black bar down the **Edit** menu until the word **Cut** is highlighted.

```
       ≰  File  Edit
                Can't Undo            ⌘z

                Cut                   ⌘H
                Copy                  ⌘c
                Paste                 ⌘v
                Clear                 ⌘H
                Paste and Discard
                Convert Clipboard...  ⌘m

                Select All Cells      ⌘A

                Nesting               ▶

                Preferences           ▶
                ✓Long Menus
```

Release the button. The cell you selected should now be cut from your notebook.

There is a danger involved in cutting cells: what if you cut the wrong cell by mistake? No matter. Pull down the **Edit** menu again and release on **Paste** and your cell will reappear.

Take a minute to practice a **Cut** and **Paste**.

SCROLLING THROUGH YOUR NOTEBOOK

Perhaps you would like to look at an earlier portion of your notebook which has by now disappeared off the top of the screen. The Macintosh folks have, of course, thought of this contingency and have provided an easy way for you to scroll through your notebook. The far right border of the notebook window is the vertical *scroll bar*. It has an up arrow at the top.

Point to this arrow and hold down the mouse button. Scroll backwards until you reach the portion of the notebook you wish to review; release the mouse button. If you want to start typing in one of the early cells, you can. All you have to do is use the mouse to position the cursor at the place where you want to type, click the mouse button, and start typing.

To scroll forward in your notebook point to the down arrow at the bottom of the

vertical scroll bar, press and hold the mouse button, and release when you've reached the desired section in the notebook. If you wish to start typing at a certain location, click at that point and begin typing.

The horizontal strip at the bottom of the notebook window lets you move in your notebook to the right and left.

Ordinarily, you won't need to scroll horizontally, so most of the time you can ignore the horizontal scroll bar. On occasion, however, you will accidentally shift your screen to the right so that part, or sometimes all, of your work disappears. This can be unnerving to say the least. It is important to remain calm. All you have to do to rectify the situation is point to the small square in the horizontal scroll bar,

click and hold on it and drag the square all the way to the left. Your work will reappear.

Practice Exercise 1: Divide the sum of 15 and 31 by the product of 2 and 5 and raise the result to the power three-quarters displaying the final answer using 9 digits. (*Ans.* 3.14100316) If you don't get this answer, practice editing your input cell until the output is correct. When you are done, Cut all of the cells you used.

Using *Mathematica* to Do High School Algebra

FACTOR AND EXPAND

Mathematica makes very quick work of high school algebra problems. For example, you know that $2x^2 + 5x - 3$ factors into $(x + 3)(2x - 1)$. The **Factor** command finds the factors of polynomial expressions. Type and *enter* **Factor[2x^2 + 5x - 3]**. Remember to use square brackets and a capital **F**.

```
In[12]:=
        Factor[2x^2 + 5x - 3]

Out[12]=
        (3 + x) (-1 + 2 x)
```

Notice that *Mathematica* writes the polynomial factors in ascending powers of x. This is typical of *Mathematica* output, and, although it looks a bit peculiar at first, you will get used to it.

You can use **Expand** to multiply algebraic expressions.

```
In[13]:=
        Expand[(3 + x)(-1 + 2x)]

Out[13]=
        -3 + 5 x + 2 x²
```

SOLVE

Give *Mathematica* a harder algebra problem. Ask it to solve the third degree polynomial equation $x^3 - 2x^2 + 9x - 18 = 0$ for x. When you type in the **Solve** command below be sure to use a double equals as shown. There is a very good reason for the double equals which needn't be gotten into right now. Have faith; in a later exploration we'll explain the difference between = and ==. The , **x** (comma x) is also crucial; it tells *Mathematica* that you want to solve for x. Don't forget to press *enter*.

```
In[14]:=
        Solve[x^3 - 2x^2 + 9x - 18 == 0, x]

Out[14]=
        {{x -> 2}, {x -> 3 I}, {x -> -3 I}}
```

Wow! You must admit that was pretty fast. Remember how such a problem was solved in precalculus—a long and tedious process of testing possible rational roots using synthetic division.

Admittedly, the output looks a little strange, but it does tell you (with no effort on your part except that of getting the command typed in properly) that the polynomial has three solutions, two are the complex numbers $3i$ and $-3i$, and the remaining one is the integer

2. Don't worry about the strange arrows and braces. There will be time enough later to consider these details.

Here is an example of the **Solve** command used to obtain the solution to the system of linear equations, $2x + 3y = 5$ and $2x - 7y = 9$, for x and y.

```
In[15]:=
        Solve[{2x + 3y == 5, 2x - 7y == 9}, {x,y}]
```

```
Out[15]=
                31        2
        {{x ->  ──,  y ->─ (-)}}
                10        5
```

You can see the solution in decimal form by typing and entering **N[%]**. In *Mathematica* the symbol **%** refers to the previous output.

```
In[16]:=
        N[%]
```

```
Out[16]=
        {{x -> 3.1, y -> -0.4}}
```

Practice Exercise 2: Find all solutions to the equation

$$6x^3 + 7x^2 - 94x + 105 = 0$$

in decimal form. (*Ans.* 1.5, 2.33333, –5)

GRAPHING

Mathematica's **Plot** command is used to draw curves in the xy-plane. Type in the command below, which will graph the equation $y = x^2 - x - 2$ for $-3 \leq x \leq 3$.

```
In[17]:=
        Plot[x^2 - x - 2, {x,-3,3}]
```

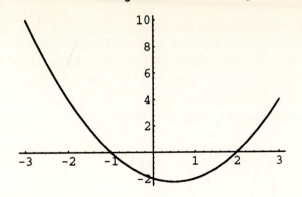

Out [17] =
        -Graphics-

If you do not get this result, read your plot command carefully. Is the **P** capitalized? Are the commas positioned correctly? The square brackets and braces must be closed. Did you type parentheses instead? Is there something in the cell other than the **Plot** command? If so, delete it or create a new cell and retype the command.

    Mathematica can draw several graphs on the same set of axes. For example, the next **Plot** command graphs the two lines $y = -2x + 5$ and $y = 3x - 1$ on the interval $-5 \le x \le 5$. Notice that the expressions for the two lines are enclosed in braces.

In[18]:=
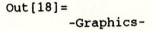

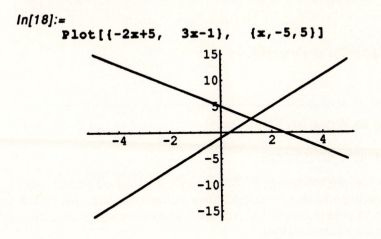

Out [18] =
        -Graphics-

Following is the trigonometric plot, $y = \sin(2x)$ for $-2\pi \le x \le 2\pi$. Be careful when you translate these expressions to the **Plot** command. In *Mathematica*, sin is written **Sin**; its argument, **2x**, is enclosed in square brackets *not* in parentheses; and you must write **Pi** for π.

```
In[19]:=
        Plot[Sin[2x], {x,-2Pi,2Pi}]
```

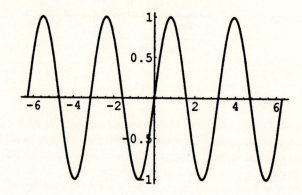

```
Out[19]=
        -Graphics-
```

Practice Exercise 3: Plot $y = \cos(3x)$, $0 \le x \le \pi$.

Preparing to Work on Your Own

OPENING A NEW NOTEBOOK

Now that you have worked through all of the example problems and practice exercises, you are ready to begin the assignment. Open a new notebook to contain your work. Point to the word **File** in the main menu bar; press and hold the mouse button. Drag down and release the mouse on the word **New**.

An untitled notebook will appear. You should give this new notebook a title and save it onto a 3.5" floppy disk. The instructions for this are given below. If your disk is a new one, it must first be initialized.

INITIALIZING A DISK

Insert your new 3.5" floppy disk, metal end forward and label up, in the slot located to the right and below the monitor. A dialog box will appear. If a dialog box does not appear, it means that your disk has already been initialized. Skip to the next section.

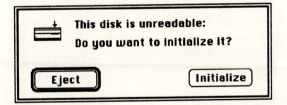

The dialog box will ask if you wish to initialize the disk. Your answer is yes. Click on **Initialize**. (If you click on **Eject**, the computer will, as you might guess, eject the disk.) If you have a different dialog box, click on the word which indicates that you wish to initialize the disk. The computer will then ask you a series of questions. The exact nature and sequence of questions varies from one type of Macintosh to another. You will be warned that initialization will erase any existing files on the disk. Since your disk is new and there are no files stored on it, it is OK to erase all files, so you can click on the word that permits erasure. You will be asked to name the disk. At that point you type in a name for the disk and click "OK." The initialization process takes a few minutes. Your Macintosh will flash messages to you such as "formatting disk" or "verifying format" or "creating directory" which let you know that it is busy, and it will tell you when the initialization is complete. Be patient.

You are now ready to give your notebook a title and save it on the floppy disk.

SAVING NOTEBOOKS

You will eventually have many *Mathematica* notebooks full of your investigations and explorations of calculus problems and ideas, so it is important to give each of them a different title. The title should be short but contain enough information so that you can refer back to it later and have some notion of the notebook's contents. For example, you might title this new notebook "your name–Assignment 1." Here's how to go about the naming. Point to the word `File` in the main menu bar. Press and hold the mouse button and drag to the phrase `Save As`. Release the mouse button. A dialog box will appear.

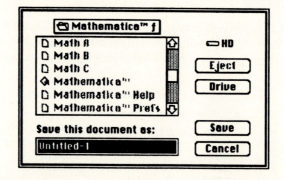

Take a minute to look at the various features of the box. There is a column of rectangles on the right side of the box labeled `Eject`, `Drive`, `Save`, and `Cancel`. These rectangles are referred to as *buttons* in Macintosh-ese. Use the mouse to point to the `Drive` button and click the mouse button a time or two observing how the contents of the dialog box change with the clicks. Pay particular attention to the disk icon at the top of the column of buttons. It will change back and forth from the rectangular hard drive icon to the square floppy disk icon.

⊂⊃ HD

▣ (disk name)

If the hard disk icon is present when you save your file, your notebook will be recorded on the computer's internal hard disk drive. You don't want this to happen. You want to carry your notebook out of the room with you, which means you must save it to the floppy disk. So, **make sure the floppy disk icon is at the top of the column of buttons before going on.** If the hard drive icon is there, click on the drive button. Now you should see the floppy icon.

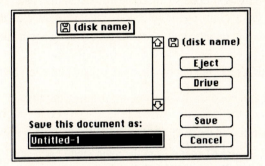

Type "your name–Assignment 1" or whatever name you wish your notebook to have. The name will be recorded in the highlighted rectangle at the bottom left of the dialog box.

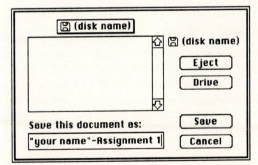

Point to the **Save** button in the dialog box and click.

The dialog box will disappear from the screen and you will hear a grinding noise as the computer records your notebook on the floppy disk. When this process is complete, your notebook window will be displayed with its new title written on the title bar. You are now ready to do the assignment in the new notebook that you just named and saved; any work already done in your notebook is now saved on the disk. However, if you continue, your new work will not be saved automatically. You must request that it be done. To do this, pull down the **File** menu as before, but this time release on **Save**. The current version of your notebook will *replace* the disk file you created with your initial save.

Exiting a *Mathematica* Session and Beginning the Next

EXITING *MATHEMATICA*

When you have finished the problems, or run out of time, you need to save your assignment notebook, quit *Mathematica*, and turn off the computer. Follow these steps:

- If you have not already done so, insert an initialized disk and save your work as described above.
- When the save is complete, pull down the **File** menu again and release on **Quit**.
- Close all windows by clicking on the small square at the left end of each title bar.

  When you are finished there should be no windows open on the screen.
- Pull down the **Special** menu from the main menu bar and release on **Shut Down**. Your disk will be ejected and the computer will turn itself off.
- Paste a label on your floppy disk and write "Assignment 1" on it so that you know what the disk contains.

REOPENING A NOTEBOOK SAVED ON A FLOPPY DISK

If you have saved a notebook, then when you go back to the computer you can continue working in the same notebook. Or, if you prefer, you can start working in a new notebook. Here are the instructions for using a notebook you have saved. Turn on the computer. Wait a half minute or so while you are being welcomed to the Macintosh. Then insert the floppy disk containing the notebook you wish to reopen into the disk slot and double click on the floppy disk icon when it appears in the upper right-hand corner of the screen.

(disk name)

A window will appear. Double click on the icon of the notebook you wish to use.

```
┌────────────────────────────────────────────┐
│ ▤□▤▤▤▤▤▤ (disk name) ▤▤▤▤▤▤▤ ▣□▣ │
├────────────────────────────────────────────┤
│  1 item        22K in disk   1,394K available│
│                                          ⬆  │
│            ┌─────┐                           │
│            │ ✿   │                           │
│            └─────┘                           │
│      "your name"-Assignment 1                │
│                                          ⬇  │
│ ◁                              ▷▣ │
└────────────────────────────────────────────┘
```

Your notebook will open. Scroll to the location in the notebook where you want to work. Use the mouse to position the cursor in a cell at the point where you wish to resume typing and click the mouse. Or, if you want to begin work in a new cell, create one and start typing.

Code for Practice Exercises

1. `N[((15 + 31)/(2*5))^(3/4)]`
2. `Solve[6x^3 + 7x^2 - 94x + 105 == 0, x];`
   `N[%]`
3. `Plot[Cos[3x], {x,0,Pi}]`

Problems

Solve each of the following problems in *Mathematica*. If your instructor wants you to hand in written solutions, use a pencil (or pen) and paper to record the input and output cells which provide the solution. If there is a printer attached to your Macintosh and you are certain you know how to use it, by all means print out your solutions. Printers may be a mystery to you. Don't fret. Use pencil and paper for this one assignment. The next section of this book provides detailed instructions for obtaining hard copy of *Mathematica* notebooks from a printer. It may be that your instructor wants you to hand in the floppy disk on which your assignment is saved; in that case you don't have to worry about pencils, pens, paper, or printers.

1. Find decimal solutions to the following arithmetic problems:
   (a) Multiply the sum of 7 and negative 12 by the third power of 2.
   (b) Add the fifth root of 4 to the reciprocal of negative π.

2. Solve each of the following polynomial equations for x. Obtain the solutions in decimal form.

   (a) $3x^2 - 5x - 28 = 0$
   (b) $9x^3 + 60x^2 - 27x - 42 = 0$
   (c) $x^4 - 4x^3 - x^2 - 16x - 20 = 0$

3. Solve the following systems of equations for x and y. Obtain the solutions in decimal form.

   (a) $13x - 4y = 5$ and $8x - 11y = 2$
   (b) $x^2 + y^2 = 25$ and $x - 2y = 1$

4. Plot each of the following curves.

   (a) $y = 3x^2 - 5x - 28, -3 \le x \le 5$
   (b) $y = x^4 - 4x^3 - x^2 - 16x - 20, -2 \le x \le 6$
   (c) $y = 1/(x^2 - 4), -3 \le x \le 3$
   (d) $y = -3\sin(x/2), 0 \le x \le 2\pi$

5. Use Mathematica's **Expand** command to multiply out the following:

   (a) $(x - 2)(x + 5)(x - 7)$
   (b) $(x + 2y)^3$
   (c) $(2a - 5b - c)^2(a + c)^5$

6. The **Together** command adds fractions. Use it to add $2/(x + y)$ to $5/(x - y)^2$ by entering the command

$$\text{Together}[2/(x + y) + 5/(x - y)^2].$$

7. Using the **Plot** command, graph the line $3x + 4y = 7$ and three other lines parallel to it on the same set of coordinate axes. Sketch the result, labeling each line with its equation.

8. Find a polynomial equation that has the solutions $x = 1, 2, 3, 4, 5$. Verify your work using **Solve**.

2

Using Graphs and Tables

The Box Problem

A open-top box is to be made from a 6 ft by 8 ft piece of cardboard by cutting out squares of equal size from the four corners as shown below and folding up the flaps on the dashed lines to form sides.

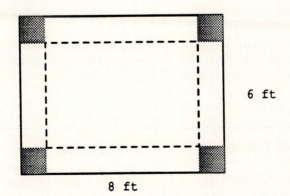

6 ft

8 ft

How large should the cut-out squares be in order to make the box with the largest possible volume?

Suppose we use the letter x to stand for the length of the sides of the cut-out corner squares. Then the dimensions of the box which we get by bending up the flaps can be expressed as follows:

$$\text{Length} = 8 - 2x \text{ feet, Width} = 6 - 2x \text{ feet, Height} = x \text{ feet.}$$

The box's volume will be the product of these three dimensions:

$$\text{Volume} = (8 - 2x)(6 - 2x)x$$

In these formulas, x is a variable whose possible values are subject to physical restrictions. For example, since x represents an actual physical length, x cannot have a negative value. Also, since the piece of cardboard is only 6 feet wide, it would be impossible to cut out corner squares having sides more than 3 feet long. These restrictions imply that x must have a value between 0 and 3.

With these observations in mind and using the symbol $V(x)$ to stand for the volume of the box that results from a corner cut of length x, we can reformulate the box problem as follows:

> Find the value for x which produces the largest value for the function $V(x)$, where $V(x) = (8 - 2x)(6 - 2x)x$ and where the values of x are restricted to the interval $0 \le x \le 3$ feet.

The largest and smallest values of a function are called its *optimum values*, so the type of question posed here is usually referred to as an *optimization* problem. Such problems can be approached in a number of different ways, and you may already have an idea of how to get started. If so, pretend for the time being that you are clueless and look at two different approaches which *Mathematica* facilitates. As you work through them, you will learn quite a bit about *Mathematica*'s **Table**, **TableForm**, **ListPlot**, and **Plot** commands. You will learn how to create, use, and **Clear** functions and variables, how to edit input cells using **Copy** and **Paste**, and how to structure **Plot** commands to zoom in on a graph's optimum points.

Begin by inserting a floppy disk, opening a *Mathematica* notebook, and naming and saving the notebook to the floppy disk. If you are uncertain of how these things are done, see Exploration One for instructions.

Method One: Computing a Table of Values

You should first teach *Mathematica* the rule for the volume function. This is accomplished by typing and entering the command below:

$$V[x\_] := (8 - 2x)(6 - 2x)x$$

Be careful. Notice that the **x** on the left side of the command is followed by an underscore (**_**) and that the symbol **x_** is enclosed in square brackets. The colon (**:**) preceding the equal sign is *not* a typo. It must be there. These conventions may seem peculiar at first, but they are necessary and must be followed if *Mathematica* is to understand exactly what function you have in mind. In a later exploration you will have an opportunity to make a more careful study of *Mathematica*'s three different "equal" signs, **=**, **:=**, and **==**.

```
In[1]:=
        V[x_]:= (8 - 2x)(6 - 2x)x
```

Notice that no output cell is produced when the function definition is entered.

Test the definition by asking *Mathematica* to compute the value of $V(x)$ when $x = 0$. The result should be 0 cubic feet.

```
In[2]:=
        V[0]

Out[2]=
        0
```

Did you remember to capitalize the **V**?

Try another test. When the corner cut has length 1 foot, the resulting box will have a volume of 24 cubic feet, since $V(1) = (8 - 2*1)(6 - 2*1)*1 = 24$ cubic feet. See if *Mathematica* returns 24 for **V[1]**.

```
In[3]:=
        V[1]

Out[3]=
        24
```

It does. Good!

Whenever you define a function in *Mathematica* it is always a good idea to test it, as was done here, using values of the argument for which you know the correct output. Typos are all too easy to make and, if the error is not caught at the beginning, all of the subsequent work will be flawed.

Practice Exercise 1: Calculate the value of V for $x = 0.5, 1.5, 2.0$, and 2.5.

Let's generate a table of volume values. The **Table** command makes very short work of this task. For example, suppose you wish to know $V(x)$ for $x = 0, 0.5, 1.0, 1.5, 2.0, 2.5$, and 3.0 feet. Type and enter the command

```
In[4]:=
        Table[{x,V[x]}, {x,0,3.0,0.5}]

Out[4]=
        {{0, 0}, {0.5, 17.5}, {1., 24.}, {1.5, 22.5},
          {2., 16.}, {2.5, 7.5}, {3., 0.}}
```

Slick. Not much work for a lot of information. For example, look at the second point in the table {0.5, 17.5}. It tells you that a corner cut of length 0.5 feet will produce a box with a volume of 17.5 cubic feet. Looking at all of the entries in the table you see that as x increases from 0 to 3.0 feet, the volume starts at 0, increases for a while, and then decreases back to 0 again. The optimum table value for $V(x)$ is easily picked out: 24 cubic feet, at $x = 1$ foot. Of course, the problem still isn't completely solved, since there might be larger V values for x-values not included in this particular table (and it turns out that there are indeed larger values, as we will see). The number picked out from this particular table is an estimate of the true optimum value.

Look carefully at the command that produced the table and think about what the various parts of it mean. The **{x, V[x]}** informs *Mathematica* that you want to find *pairs* of numbers $\{x, V(x)\}$. The second part of the command **{x, 0, 2.5, 0.5}** specifies which values of x to use. It says: Start the value of x at 0, end at 3.0, and increase x in steps of 0.5.

It is quite easy to change the table command so that it will generate more points. That way we can get a better estimate of maximum V.

Edit your previous **Table** command to produce the following one, then enter it.

```
In[5]:=
        Table[{x,V[x]}, {x,0,3,0.25}]

Out[5]=
        {{0, 0}, {0.25, 10.3125}, {0.5, 17.5},
          {0.75, 21.9375}, {1., 24.},
          {1.25, 24.0625}, {1.5, 22.5},
          {1.75, 19.6875}, {2., 16.},
          {2.25, 11.8125}, {2.5, 7.5},
          {2.75, 3.4375}, {3., 0.}}
```

You have the larger table before you, but in its current form it is a little hard to read. A really big table containing lots of rows would be impossible to take in. Fortunately, *Mathematica* has commands which help organize and interpret table data. Two of them, **TableForm**, which writes a table vertically, and **ListPlot**, which graphs the table as a set of points, are introduced below.

Before getting into the details of these commands, assign the large table the name **t**. Don't groan. This is very easy to do and worth the trouble. Insert the symbols **t** = before the word **Table** in the previous **Table** command and reenter the cell. Note that this time you type just = (equals) and not : = (colon equals).

```
In[6]:=
        t = Table[{x,V[x]}, {x,0,3,0.25}]
```

```
Out[6]=
        {{0, 0}, {0.25, 10.3125}, {0.5, 17.5},
         {0.75, 21.9375}, {1., 24.},
         {1.25, 24.0625}, {1.5, 22.5},
         {1.75, 19.6875}, {2., 16.},
         {2.25, 11.8125}, {2.5, 7.5},
         {2.75, 3.4375}, {3., 0.}}
```

From now on whenever you type the letter **t**, *Mathematica* will understand that you mean the entire table. Try it. Type and enter the letter **t**. You should get the large table back.

Now, ask *Mathematica* to display the table **t** vertically.

In[7]:=
```
        TableForm[t]
```

```
Out[7]//TableForm=
        0           0
        0.25        10.3125
        0.5         17.5
        0.75        21.9375
        1.          24.
        1.25        24.0625
        1.5         22.5
        1.75        19.6875
        2.          16.
        2.25        11.8125
        2.5         7.5
        2.75        3.4375
        3.          0.
```

That is much better. Scroll up and down through the table and pick out the largest table entry for $V(x)$. It is 24.0625 cubic feet and it occurs when $x = 1.25$ feet.

If you ask properly, *Mathematica* will plot the table for you. The command you need is **ListPlot**.

```
In[8]:=
        ListPlot[t]
```

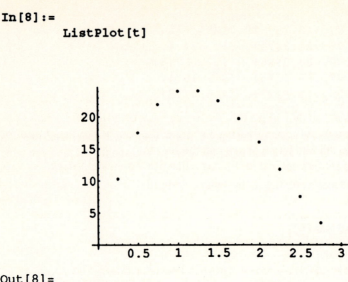

```
Out[8]=
        -Graphics-
```

Isn't that a useful picture?! *Mathematica* has plotted x on the horizontal axis, in the usual way, and the corresponding values of V on the vertical axis. The graph gives a very clear picture of how the values of $V(x)$ change as x increases from 0 to 3.0 feet.

Practice Exercise 2: Use a step size of 0.1 to estimate the maximum for $V(x)$. Make a vertical table and plot the points. What is your estimated maximum value for the volume? What size square should be cut out? (*Ans.* The maximum volume is 24.244 cubic feet, at $x = 1.1$ feet.)

This completes the work with **Table** for the time being. Below we will look at a way in which **Plot** can be used to solve optimization problems, but first we explain how to clear the definitions and assignments that you no longer need.

Clearing Definitions and Assignments

When you are through using a function or a variable, it is a good idea to remove it from *Mathematica*'s memory. Old definitions that you have forgotten about can cause strange errors that are very hard to find. Use the **Clear** command.

```
In[9]:=
        Clear[t]
```

Once the **Clear** command is entered, the letter **t** no longer has an assigned value, and it will have no special meaning until you once again set it equal to some quantity or expression. If you enter the symbol **t** now, *Mathematica* will return "t", meaning that no assignment is currently made to the letter.

```
In[10]:=
        t
```

```
Out[10]=
        t
```

Make a habit of clearing when you are done with a variable or a function. Otherwise, there is the danger that you may forget that a letter has a meaning and use it inappropriately. *Mathematica* will not forget (barring electronic catastrophe) an assignment or definition unless you tell it to forget or you quit the session.

We turn now to another method for finding a function's optimum values.

Method Two: Estimating from a Plot

Return to the Box Problem, but this time there is a tougher standard: we want the value of x which produces the maximum V accurately to two decimal places.

Begin by graphing $V(x)$ and labeling the graph's axes. The symbol -> requires two keystrokes, the hyphen followed by the greater than sign.

```
In[11]:=
        Plot[
          V[x], {x,0,3.0},
          AxesLabel->{"x","V"}
          ];
```

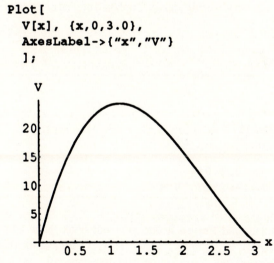

Note: The semicolon at the end of the **Plot** command suppresses the output cell. Enter the command without the semicolon, and you will see the difference. Why clutter up your notebook with output cells when all you really need is the graph? Of course, if you wish to see the output cell, leave off the semicolon. Suit yourself.

Clearly, the largest volume is the V coordinate at the high point on the graph. The first coordinate of that point is the value of x that produces the largest volume. From the graph we see that the x value of the highest point is between 1.0 and 1.5 feet. To figure out just where x is in this range, we need to zoom in on the point to get a closer view. A new cell editing technique is very helpful in doing this.

Copying a Cell

Scroll backwards until you find the cell which contains the command `Plot[V[x], {x, 0, 3.0}, AxesLabel->{"x", "V"}]`. *Select* this cell by clicking on its cell bracket. The bracket should fill with a black rectangle. Now pull down the **Edit** menu from the main menu bar and release the mouse on the word **Copy**. *Mathematica* will save a copy of that `Plot` command in a special memory area called the *clipboard*.

Move the cursor down the screen until it becomes horizontal, and click. You are now ready to paste a copy of the `Plot` command from the Clipboard into your notebook at this point. Pull down the **Edit** menu and release on **Paste**. A copy of the `Plot` command will appear in a new cell.

Now here is some really good news: The Clipboard still contains the `Plot` command, so you can make another copy of it if you want to: Just move the cursor until its position is again horizontal, and click; then select **Paste** from the **Edit** menu. There it is again *and* it is still on the Clipboard and will remain there until the next time you execute **Copy**. We'll make very effective use of Clipboard copies in the next section, on zoom-in procedures.

Zooming In

Make sure that the command

```
Plot[
  V[x], {x,0,3.0},
  AxesLabel->{"x","V"}
  ]
```

is copied to the Clipboard, and **Paste** it into your notebook. Now edit the pasted command, changing `{x, 0, 3.0}` to `{x, 1.0, 1.5}`. One way to do this is to place the cursor before the `0`, press the mouse button, and drag across `0, 3.0` so that it is highlighted. Release the mouse button and type `1.0, 1.5`. The new text will replace the highlighted text. Press the *enter* key.

```
In[12]:=
    Plot[
      V[x], {x,1.0,1.5},
      AxesLabel->{"x","V"}
      ];
```

Exploration Two

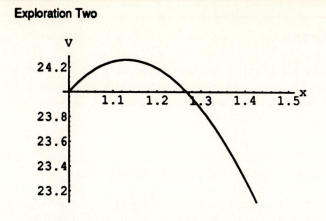

You see, you get a much closer look at the high point. Clearly the *x*-coordinate of the maximum point is between 1.1 and 1.2 feet. The value of *V* is about 24.2 cubic feet.

Note that the axes on the graph do *not* pass through the origin. They intersect at the point (1.0, 24.0). *Mathematica* tries to plot as much detail as possible in the requested region. This means that it frequently will not display the point (0, 0). Since you are probably accustomed to having the axes pass through the origin, you need to check the horizontal and vertical scales carefully whenever a *Mathematica* plot is drawn.

Zoom in again: **Paste** the **Plot** command into your notebook. This time change the *x*-range to **1.1, 1.2**. (By the way, **Paste** can be executed from the keyboard by holding down the key and pressing the v-key.)

In[13]:=
```
Plot[
    V[x],    {x,1.1,1.2},
    AxesLabel->{"x","V"}
    ];
```

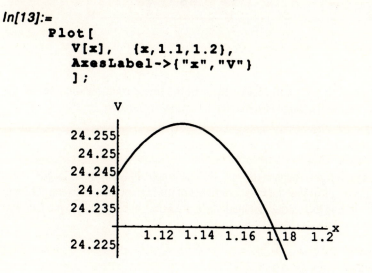

Enlarging a Graph

If the graph you get is a little hard to read because the numbers run together below the x-axis, make it larger. Use the mouse to move the cursor into the cell containing the graph. The cursor will become a small circle containing a cross.

Click on the graph. A rectangle will appear around it.

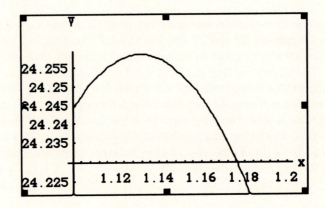

Move the cursor to the small black square in the lower right-hand corner of the rectangle. The cursor will change to a diagonal, two-headed arrow.

Drag down and to the right. Release the mouse. The graph should now be larger.

This graph makes clear that the first coordinate of the high point is between 1.12 and 1.14. We still don't have the second decimal place. **Paste** and edit again to pin it down.

```
In[14]:=
    Plot[
      V[x], {x,1.12,1.14},
      AxesLabel->{"x","V"}
      ];
```

Exploration Two

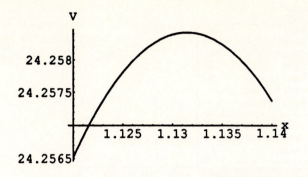

It looks like the *x* that gives the maximum volume is 1.13, to two decimal places. The corresponding volume is V(1.13):

```
In[15]:=
        V[1.13]

Out[15]=
        24.2584
```

So our estimate of the maximum volume is 24.2584 cubic feet. Since the *x* value is accurate only to two decimal places, our estimate of *V* is not likely to be any more accurate, so we should round to two decimal places. Maximum *V* is 24.26 cubic feet.

Practice Exercise 3: Find the third and fourth decimal places in the estimate for *x*, and the corresponding values for *V*. (*Ans. x* = 1.1348 feet)

Printing Your Work

Since there are probably a number of others using the same printer you are, you had better set things up so that when your notebook is printed its pages are numbered and the name of your file appears on each page. This is not difficult to do. Pull down the **File** menu to the phrase **Printing Settings** and slide the mouse to the right and down. Release on **Headers and Footers....**

```
┌─────────────────────────────────┐
│ File                            │
│  New                       ⌘n   │
│  Open...                   ⌘o   │
│  Close                     ⌘w   │
│  Save                      ⌘s   │
│  Save As...                ⌘S   │
│  Save As Other...               │
│  Open Selection                 │
│─────────────────────────────────│
│  Show Keywords                  │
│ ✓Show In/Out Names              │
│  Colors...                      │
│─────────────────────────────────│
│  Printing Settings    ▶┌─────────────────────┐
│  Print...            ⌘p│ Show Page Breaks    │
│  Print Selection...    │─────────────────────│
│─────────────────────── │ Page Setup...       │
│  Quit                ⌘q│ Printing Options... │
└────────────────────────│ Headers and Footers...│
                         └─────────────────────┘
```

A dialog box like the following one will appear.

```
┌─────────────────────────────────────────────────────┐
│          Headers/Footers for Exp 2                   │
│                                                      │
│  Starting page number: [1      ]        ┌────────┐  │
│                                         │   OK   │  │
│  ☐ No headers or footers on first page  └────────┘  │
│  ☐ Left and right pages                 ┌────────┐  │
│                                         │ Apply  │  │
│  Put the first page on the:             └────────┘  │
│     ○ Left    ○ Right      Include bar: ┌────────┐  │
│                                         │Defaults│  │
│     header: [\f "your name"       ] ☐   └────────┘  │
│                                         ┌────────┐  │
│     footer: [\p                   ] ☐   │  Help  │  │
│                                         └────────┘  │
│  Left header: [                   ] ☐   ┌────────┐  │
│                                         │ Cancel │  │
│  Left footer: [                   ] ☐   └────────┘  │
│                                                      │
│  Special objects: \p = page number, \f = filename,   │
│  \t = time, \d = date, \k = cell keywords            │
└─────────────────────────────────────────────────────┘
```

Click in the header box and type \f (for filename). Press the space bar and type your name. In the footer box type \p (for page number). Now, when your notebook is printed out, the pages will be numbered and your name and your filename will appear on each page. If you have the time and the interest you might experiment a little with the settings in this dialog box. You can add the date to the header if you like or the time of day. A decorative line can be placed at the top of each page, and so forth. When you are finished with this box, click on the **OK** button.

Always save your notebook before printing any part of it!!! If you haven't done so already, use the **Save As** command in the **File** menu to title your notebook and save it to your floppy disk. See Exploration One for instructions. Now you are ready to obtain a printed copy of the cells in your *Mathematica* notebook for which you need a written record. Here's how it is done: Select the cell or cells you want printed by clicking on their cell brackets. When more than one cell is to be printed, hold down the ⌘-key as you click. Skip the cells which you don't want to print. After all of the cells you need printed are selected, pull down the **File** menu and release on **Print Selection**.

```
 🍎  File
        New              ⌘n
        Open...           ⌘o
        Close            ⌘w
        Save             ⌘s
        Save As...        ⌘S
        Save As Other...
        Open Selection

        Show Keywords
      ✓ Show In/Out Names
        Colors...

        Printing Settings   ▶
        Print...          ⌘p
        Print Selection...

        Quit             ⌘q
```

A dialog box will appear on which you can indicate the number of copies you wish printed.

```
 LaserWriter  "LaserWriter II NT"                    5.2        [   OK   ]

 Copies: [1]        Pages: ⦿ All  ○ From: [    ]  To: [    ]    [ Cancel ]

 Cover Page:   ⦿ No ○ First Page ○ Last Page                    [  Help  ]

 Paper Source: ⦿ Paper Cassette ○ Manual Feed
```

Click on the **OK** button.

To print the whole notebook pull down the **File** menu and release on **Print**. Your paper copy can then be obtained from the printer connected to the Macintosh you are using.

Two Important Points to Remember

SAVE YOUR WORK

Open a new notebook for your assignment. As you work, remember to **save your notebook periodically to a floppy disk**. Every 15 minutes or so, stop and save your file. The first time you save, use **Save As** from the file menu. Then you can save from the keyboard by holding down the ⌘-key and pressing **s**, if you wish. If an accident occurs and your notebook or part of it is trashed for some reason, you will not lose more than 15 minutes of work. Most computer users learn the cruel, hard way from the sickening loss of a big notebook in some sudden electronic glitch that files must be regularly saved to a floppy disk. Profit from the sad experience of *Mathematica* veterans: **Save your notebooks regularly.**

REENTER DEFINITIONS

When you open a notebook that you previously saved on a floppy disk, you must reenter any cell whose output you want to use. For example, if you wish to use a function you defined in an earlier session and saved in the notebook, you must find that function, click on its cell bracket, and press the *enter* key. In other words, you must reteach *Mathematica* the function.

Code for the Practice Exercises

```
1. V[0.5],   V[1.5],   V[2.0],   V[2.5]
2. Plot[
     V[x], {x,1.13148,1.13149},
     AxesLabel->{"x","V"}
     ];
```

Problems

1. Consider the function $f(x) = 9x^4 - 36x^3 - 21x^2 + 25x + 18, \ -1 \leq x \leq 4$.
   (a) Print out a table of function values, written vertically, with a step size of 0.5. Use it to estimate the function's maximum and minimum values.
   (b) Use **ListPlot** to graph the table; print the resulting plot.
2. **Plot** the function in Problem 1. Zoom in on the graph to obtain the coordinates of the high point accurately to two decimal places. Print the final plot together with your estimated maximum function value and the corresponding value for *x*.
3. Repeat the instructions in Problem 2 for the graph's low point.

4. An open-top box is to be made from a 8-ft by 8-ft piece of cardboard by cutting out squares of equal size as shown below and folding on the dashed lines to form sides.

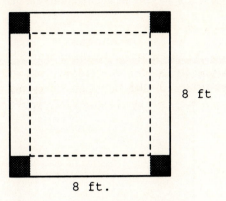

8 ft

8 ft.

(a) If the cut-away squares have sides of length 1-foot, what are the dimensions of the box? What is the volume of the resulting box?

(b) Letting x stand for the length of the side of a cut out corner square, write a formula for the volume V in terms of x.

(c) Write an inequality which expresses the physical restrictions on x.

(d) Use *Mathematica* to generate a table of 20 values of V for x-values in the interval found in (c). Obtain the table in vertical form and use it to estimate the largest possible volume for the box.

(e) Use **Plot** to zoom in on the graph of V until you have found the optimal x-value to three decimal places. Compute the corresponding value for V.

(f) Print the table and the final graph from (d) and (e).

5. Define a function **f[x_]:= x^2 - 4x + 3**. Type and enter the command **Table[{x, f[x]}, {x, 0, 10}]**. Note that no step size is included in this command. Write a sentence explaining what *Mathematica* does when the step size is left out.

6. Type the command **Table[2x + 1, {x, 0, 20}]**. What is the result? Write a **Table** command that gives a list of the first ten powers of 2. (That is, the output should be {2, 4, 8, 16, ...}.)

7. A rectangular truck yard is to be constructed adjoining a warehouse. The warehouse wall will form one side of the yard. The other three sides will be enclosed using 500 feet of Cyclone fencing and two gates. A 3-foot-wide pedestrian gate will be installed adjacent to the warehouse on one side of the yard and a 30-foot-wide truck gate on the opposite side.

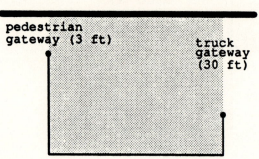

(a) If the side containing the pedestrian gate is to be 150 feet wide, what will be the area of the truck yard?
(b) Letting x stand for the length of the side containing the pedestrian gate, write a formula for the area A of the truck yard in terms of x.
(c) Write an inequality which expresses the physical restrictions on x.
(d) Print out a table in vertical form which lists values of A for at least 20 values of x.
(e) Use **Plot** to find the maximum value for A accurate to three decimal places.
(f) What are the dimensions of the lot with the maximum area?

3

The Rocket Problem

> The fuel in a certain rocket burns for 60 seconds. After t seconds, the rocket's distance h above the ground, measured in feet, is given by the formula
>
> $$h(t) = 40t^2, \ 0 \le t \le 60.$$
>
> Investigate the motion of the rocket.

Teach *Mathematica* the height function. Begin by clearing the variables you are going to use just in case there are some values assigned to them.

```
In[1]:=
        Clear[h,t]
        h[t_]:= 40t^2
```

A number of questions about the motion of the rocket can be answered without using calculus—high school algebra and arithmetic are all that is needed, with *Mathematica* doing the grunt work for you.

Question 1: How high does the rocket travel before it runs out of fuel?

The fuel runs out when $t = 60$, so $h(60)$ is the distance asked for.

```
In[2]:=
        h[60]
```

Out[2]=
>   144000

When it runs out of fuel, the rocket is 144, 000 feet above the ground.

Question 2: What is the average speed of the rocket during its first 60 seconds aloft?

Recall that the average speed of a moving body is calculated as the distance traveled divided by the time elapsed. In this case the average speed will be 144,000/60 ft/sec.

In[3]:=
>   **144000/60**

Out[3]=
>   2400

Thus the answer is that rocket travels at an average rate of speed of 2400 feet per second during its first 60 seconds aloft.

If you hadn't already calculated the distance figure, 144,000 feet, here is a way you could have computed the same average speed:

In[4]:=
>   **(h[60] - h[0])/60**

Out[4]=
>   2400

Practice Exercise 1: Calculate the average speed of the rocket over the 5-second time interval from $t = 10$ seconds to $t = 15$ seconds. (*Ans.* 1000 ft/sec)

You can get a rough idea of how the rocket moves during the first 60 seconds by constructing a table showing the altitude of the rocket at various times. The command below gives the table the name *heights*.

In[5]:=
>   **heights = Table[{t,h[t]},{t,0,60,10}]**

Out[5]=
>   {{0, 0}, {10, 4000}, {20, 16000},
>   {30, 36000}, {40, 64000}, {50, 100000},
>   {60, 144000}}

Let's view the table vertically and add a heading to it. The **TableHeadings** setting must be typed carefully. Don't leave out any of the braces or the quotation marks.

```
In[6]:=
    TableForm[
      heights,
      TableHeadings->
        {None,
        {"time\nin seconds\n",
        "height\ninfeet\n"}
        }
      ]

Out[6]//TableForm=
    time        height
    in seconds  in feet

    0           0
    10          4000
    20          16000
    30          36000
    40          64000
    50          100000
    60          144000
```

You have to admit that when the table is written vertically and labeled, it is much easier to read. It *is* a bit of a nuisance to get the **TableForm** command typed correctly, but it is worth the effort.

The setting for the **TableHeadings** option needs a word or two of explanation. The portion of it which reads

```
        {"time\nin seconds\n",
        "height\nin feet\n"}
```

causes the labels on the columns to read as they do. The symbol \n creates a new line in the heading. The setting **None** informs *Mathematica* that you want the rows of the table to go unlabeled.

Returning to the problem at hand, look at the details of the table because they do reveal a great deal about the motion of the rocket. The first row of the table says that when $t = 0$, $h = 0$. In other words, initially the rocket is on the ground. The second row tells you that at the end of its first 10 seconds aloft, the rocket is 4000 feet above the ground. So, the rocket rose 4000 feet in the first 10 seconds for an average velocity of 400 ft/sec during this time interval.

The arithmetic was very simple in this average speed calculation. If you hadn't been able to do the calculation in your head, the following command would have done the job for you.

```
In[7]:=
        h[10] - h[0])/10
```

```
Out[7]=
        400
```

Notice that the rocket moves faster during its second 10 seconds aloft. At the end of 20 seconds its elevation is 16,000 feet above the ground, 12,000 feet higher than it was at the end of the first 10 seconds. So, in the second 10 seconds the rocket rose 12,000 feet, giving an average velocity of 1200 ft/sec.

```
In[8]:=
        (h[20] - h[10])/10
```

```
Out[8]=
        1200
```

Practice Exercise 2: Calculate the average velocity of the rocket over the remaining four 10-second time intervals. (*Ans.* 2000 ft/sec, 2800 ft/sec, 3600 ft/sec, 4400 ft/sec)

All of these average velocities can be collected together into a single table and displayed vertically with the starting and ending times for the intervals over which the average velocities were computed.

```
In[9]:=
        Clear[k];
        TableForm[
          Table[
            {k-10, k, (h[k] - h[(k-10)])/10},
            {k,10,60,10}
            ],
          TableHeadings->
            {None,
            {"time\nstart\n",
            "time\nend\n",
            "avg\nvelocity\nft/sec\n"}
            }
          ]
```

```
Out[9]//TableForm=
        time      time      avg
        start     end       velocity
                            ft/sec


        0         10        400
        10        20        1200
        20        30        2000
        30        40        2800
        40        50        3600
        50        60        4400
```

Clearly, the rocket is speeding up as it rises. With that notion in mind, we are now ready to ask a question which precalculus methods alone cannot answer.

Question 3: How fast is the rocket moving when it has been aloft exactly 3 seconds?

Note that we are *not* asking here for the *average* speed of the rocket during its first 3 seconds aloft. That would be easy to compute. The answer is 120 ft/sec.

```
In[10]:=
        (h[3] - h[0])/3

Out[10]=
        120
```

We are asking a more profound kind of question. If the rocket contained an ideal speedometer which perfectly registered its exact speed, then what would that speedometer's reading be at the end of precisely 3 seconds of flight? In other words, we are asking for the rocket's *instantaneous speed* when $t = 3$. Dividing distance by time will not compute the speed at an instant, because the amount of time which elapses during an instant, say the instant when $t = 3$, is 0. Division by 0 in the speed formula would be impossible.

Even though we cannot directly calculate the rocket's instantaneous speed when $t = 3$, we can still obtain a good estimate by reasoning as follows: In a very small time interval after the instant when $t = 3$, say 0.1 second, the rocket will speed up only a little bit, so if we compute the average speed over this small time interval we will obtain a number which is close to the instantaneous speed when $t = 3$.

```
In[11]:=
        (h[3.1] - h[3])/0.1

Out[11]=
        244.
```

Since the rocket is speeding up as it rises, this number, 244 ft/sec, is slightly larger than the instantaneous speed at time 3, and is, therefore an *upper bound* for the instantaneous speed. Now, if we compute the average speed over the tenth of a second time interval before 3 seconds, we will obtain a number which is near the speed at 3 seconds but slightly smaller. It will be a *lower bound* for the instantaneous speed.

```
In[12]:=
        (h[3] - h[2.9])/0.1
```

Out[12]=
        236.

Great! Now we know that the speed at 3 seconds is between 236 ft/sec and 244 ft/sec. Refine the estimate by narrowing the time interval before and after three seconds to 1/100 of a second. (Remember: You can **Copy** and **Paste** the commands which gave the previous estimates and edit then to narrow the time interval.) The upper bound for the instantaneous speed at 3 seconds is 240.4 ft/sec, calculated as follows:

```
In[13]:=
        (h[3.01] - h[3])/0.01
```

Out[13]=
        240.4

And the lower bound is

```
In[14]:=
        (h[3] - h[2.99])/0.01
```

Out[14]=
        239.6

Using the symbol v(3) to stand for the instantaneous speed, or *velocity*, when $t = 3$, these last two calculations tell us that $239.6 \leq v(3) \leq 240.4$.
    Let's get even closer. Use a time interval of 1/1000 of a second.

```
In[15]:=
        (h[3.001] - h[3])/0.001
```

Out[15]=
        240.04

```
In[16]:=
        (h[3] - h[2.999])/0.001
```

Out[16]=
        239.96

So, now we know that 239.96 ft/sec $\leq v(3) \leq$ 240.04 ft/sec. This inequality pins down v(3) to a very small interval; the difference between the upper and lower bounds is only

0.08 ft/sec. But by taking shorter and shorter time intervals, even greater precision could be obtained. Here is a table of lower bounds for shorter time intervals:

```
In[17]:=
        Clear[k,dt,lbound];
        dt = 10.^-k;
        lbound = (h[3]-h[3-dt])/dt;
        TableForm[
          Table[{dt, N[lbound,12]}, {k,1,7}]
          ]
```

```
Out[17]//TableForm=
        0.1         236.
        0.01        239.6
        0.001       239.96
        0.0001      239.996
        0.00001     239.9996
                -6
        1. 10       239.99996
                -7
        1. 10       239.999996
```

Here is a table, fancied up with headings, which shows both upper and lower bounds for v(3):

```
In[18]:=
        Clear[k,dt,lbound,ubound];
        dt = 10.^-k;
        lbound = (h[3] - h[3 - dt])/dt;
        ubound = (h[3 + dt] - h[3])/dt;
        TableForm[
          Table[
            {dt,N[lbound,12],N[ubound,12]},
            {k,1,7}
            ],
          TableHeadings->
            {None,
            {"time\ninterval\nin sec\n",
            "v(3)\nexceeds",
            "but is less than\nin ft/sec\n"}
            }
          ]
```

```
Out[18]//TableForm=
        time          v(3)              but is less than
        interval      exceeds           in ft/sec
        in sec

        0.1           236.              244.
        0.01          239.6             240.4
        0.001         239.96            240.04
        0.0001        239.996           240.004
        0.00001       239.9996          240.0004
               -6
        1. 10         239.99996         240.00004
               -7
        1. 10         239.999996        240.000004
```

There is no mathematical reason why this table could not be continued indefinitely! Imagine doing so. The first nine rows make the pattern of the successive entries clear, and it is obvious what the value for v(3) must be. There is only one number which is larger than all the lower bounds and smaller than all the upper bounds. That number, obviously, is 240. So, $v(3) = 240$ ft/sec.

Practice Exercise 3: Find upper and lowers bounds for the velocity of the rocket when $t = 15$ using time intervals of length 0.1, 0.01, and 0.001. (*Ans.* $1196 \le v(15) \le 1204$, $1199.6 \le v(15) \le 1200.4$, $1199.96 \le v(15) \le 1200.04$)

You say there ought to be an easier way to find these instantaneous velocities? One that avoids all of these approximations? You are right. But, keep in mind that you will only be able to follow the shortcut method if you have struggled through the approximations. The shortcut is based on seeing the pattern in the form of the calculation made to obtain each of the approximations. Take the example of finding the velocity at t = 3. Each of the approximations to this velocity had essentially the same form, $(h(3 + dt) - h(3))/dt$, the average velocity of the rocket over a small time interval dt. (When a lower bound was calculated, the value of dt was replaced by $-dt$.) This average velocity got closer to the instantaneous velocity as the value of dt neared zero. By observing the average velocities calculated for dt near zero we were able to establish the figure 240 ft/sec for the instantaneous velocity at $t = 3$ seconds.

Mathematica will find this value for you, if you ask it to find the number that the fraction $(h(3 + dt) - h(3))/dt$ is approaching as the value of dt nears zero. The **Limit** command accomplishes this task.

```
In[19]:=
        Clear[dt];
        Limit[(h[3 + dt] - h[3])/dt, dt->0]
        General::load: Loading package Series'.

Out[19]=
        240
```

It might take a while for *Mathematica* to make this **Limit** calculation because before it can execute the command it first has to load a special piece of software. It is worth the wait. Once this package is loaded, subsequent **Limit** commands are executed with blinding speed. For example, ask *Mathematica* for the instantaneous velocity when $t = 15$. This is the velocity which you approximated in the second practice exercise.

```
In[20]:=
        Limit[(h[15 + dt] - h[15])/dt, dt->0]
```

```
Out[20]=
        1200
```

So, the instantaneous speed of the rocket at time $t = 15$ seconds is 1200 ft/sec.
What is the velocity of the rocket when it runs out of fuel?

```
In[21]:=
        Limit[(h[60 + dt] - h[60])/dt, dt->0]
```

```
Out[21]=
        4800
```

Why don't we just teach *Mathematica* a formula for the instantaneous velocity at time t?

```
In[22]:=
        v[t_]:=
          Limit[(h[t + dt] - h[t])/dt, dt->0]
```

Now use the formula to construct a table of the instantaneous velocities and heights for the rocket calculated at 10-second intervals.

```
In[23]:=
        Clear[t,dt,k];
        TableForm[
          Table[{k,h[k],v[k]}, {k,0,60,10}],
          TableHeadings->
            {None,
            {"time\nin sec\n",
            "height\nin ft\n",
            "velocity\nin ft/sec\n"}
            }
          ]
```

```
Out[23]//TableForm=
        time        height      velocity
        in sec      in ft       in ft/sec

        0           0           0
        10          4000        800
        20          16000       1600
        30          36000       2400
        40          64000       3200
        50          100000      4000
        60          144000      4800
```

Take a look at those velocities. It seems that each one is exactly 80 times the corresponding time value. Is this a coincidence? Of course not! No reason to bring it up otherwise. See what happens when you ask *Mathematica* to evaluate the formula for the instantaneous velocity for the rocket at time t, a general time value.

```
In[24]:=
        Limit[(h[t + dt] - h[t])/dt, dt->0]
```

```
Out[24]=
        80 t
```

There you have it. At each instant the velocity of the rocket is 80 times the elapsed time.

With this information you can paint a very clear verbal picture of the rocket's motion while it has fuel: The rocket moves straight up from the ground, and as it rises it speeds up in such a way that its velocity at each instant is 80 times the elapsed time. After 60 seconds, having reached an elevation of 144,000 feet and a speed of 4800 ft/sec, it runs out of fuel.

Before going on to the assignment, **Clear** all of the variables you've used.

```
In[25]:=
        Clear[h,v,t,dt,heights,lbound,ubound]
```

Code for the Practice Exercises

1. `(h[15] - h[10])/5`
2. `(h[30] - h[20])/10`
   `(h[40] - h[30])/10`
   `(h[50] - h[40])/10`
   `(h[60] - h[50])/10`

3. For example,

```
(h[15.001] - h[15])/0.001 and
(h[15] - h[14.999])/0.001
```

Problems

1. Construct a table of upper and lower bounds for the velocity of the rocket when $t = 30$ using time intervals of length 0.1, 0.01, 0.001, 0.0001. What value do these calculations suggest for $v(30)$? Check your work with a `Limit` command.

2. A weather balloon is moving vertically. After t hours its distance h above the ground, measured miles, is given by the formula $h(t) = 9t - 3t^2$, $0 \le t \le 3$. Analyze the motion of the weather balloon.

    (a) Print out a table which gives the position of the balloon at half-hour intervals.
    (b) Print out a table which gives the average velocity of the balloon over each half-hour interval.
    (c) Write a limit formula for the balloon's instantaneous velocity and use it to make a table of instantaneous velocities calculated at the end of each half-hour interval.
    (d) Use a `Limit` command to derive a formula for the balloon's instantaneous velocity at time t.
    (e) Write a short paragraph describing the motion of the balloon.

3. A particle starts at the origin and moves along the x-axis for exactly 5 minutes. Its position after t minutes is given by the formula $x(t) = 3t^4 - 32t^3 + 72t^2$.

    (a) Print out a table which gives the position of the particle at the end of each half-minute interval and the average velocity of the particle over each half-minute interval.
    (b) Use `Limit` commands to derive the particle's instantaneous velocity at the end of each half-minute interval.
    (c) Use a `Limit` command to derive a formula for the particle's instantaneous velocity at time t.
    (d) Write a short paragraph describing the motion of the particle.

4

Curves and Slopes

A line can be described by giving a point on it and its slope. For example, if you know that a line passes through the point (1, 2) with a slope of 3, then you can write its equation as $y = 3(x - 1) + 2$. We're going to plot this line in a minute. We *know* you already know how to graph lines, but we are going to do this one anyway because it will provide us an opportunity to tell you about a very useful **Plot** option in a familiar context.

The option is called **Prolog** and it permits you to add lines, points, text, and other features to the display of the graph. The command below makes use of the **Prolog** option to add an enlarged point to a plot of the line. Take a look at it. The **Prolog** setting asks *Mathematica* to enlarge the point (–2, 4) to a size that is 3/100 of the width of the plot. The double set of braces in the setting are a technicality, needed to make sure that the **PointSize** value applies only to the following **Point** command and not to any other part of the plot. Remember that the arrow –> is made up of a minus sign followed by a greater than symbol.

```
In[1]:=
      Clear[x,y];
      y[x_]:= 3(x - 1) + 2;
      Plot[
        y[x], {x,-2,4},
        Prolog->
          { {PointSize[0.03], Point[ {1,2} ]} }
        ];
```

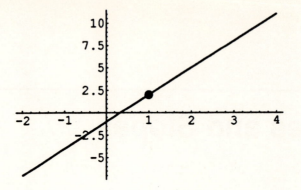

Let's take a little time to review slope calculations, since we will be making a lot of them.

You know how to calculate the slope of a line from two of its points. For example, the line above passes through the points (2, 5) and (3, 8). So, the following ratio should give the slope of the line:

```
In[2]:=
        (8 - 5)/(3 - 2)
```

```
Out[2]=
        3
```

And it does. You get the same result no matter which two points you choose. Try the slope calculation with arbitrary points $(a, y(a))$ and $(b, y(b))$.

```
In[3]:=
        (y[b] - y[a])/(b - a)
```

```
Out[3]=
        -3 (-1 + a) + 3 (-1 + b)
        ───────────────────────
                -a + b
```

Simplify the slope expression.

```
In[4]:=
        Simplify[%]
```

```
Out[4]=
        3
```

Every single a and b cancelled out! That means that no matter what values these variables have in the original slope expression, when the arithmetic is finished, the final slope ratio will be 3. The fact that the slope of a line does not depend on which two points you choose to calculate it is a special characteristic of straight lines. For a curve that is not straight, the slope *does* depend on the points chosen. For example, consider $y = f(x) = x^2$.

```
In[5]:=
        Clear[f,x];
        f[x_]:= x^2
```

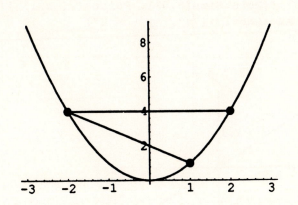

The slope from (–2, 4) to (2, 4) is 0.

```
In[6]:=
        (f[2] - f[-2])/(2 - (-2))
```

```
Out[6]=
        0
```

But the slope from (–2, 4) to (1, 1) is –1.

```
In[7]:=
        (f[1] - f[-2])/(1 - (-2))
```

```
Out[7]=
        -1
```

The slopes of the two segments are obviously quite different. Straight lines have the same slope over every segment of their graph, curves do not.

The Slope of a Curve at a Point

This doesn't mean that the notion of slope cannot be applied to curves. It can, and in this section we will investigate a method for doing so. Look at the plot below, which shows the graph of $f(x) = x^2$ in a small region near the point (–1, 1).

```
In[8]:=
        Plot [
          f[x],{x,-1.1,-0.9},
          Prolog->
            {{PointSize[0.03], Point [ {-1,1} ]}},
          AxesLabel->{"x", "y = x^2"}
          ];
```

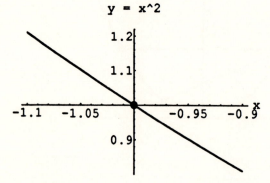

As you study this plot, keep in mind that the axes intersect at (–1, 1) not at the origin, and that the scale on the axes is quite small. It is as if we had put the curve $f(x) = x^2$ under a microscope with the cross hairs centered at (–1, 1) and so what we have in view is a greatly enlarged segment of the curve. Notice that it is very nearly straight. The slopes across parts of this segment will not differ from one another by very much. For example, calculate the slope between the points (–1.1, f(–1.1)) and (–1.0, f(–1.0)).

```
In[9]:=
        (f[-1] - f[-1.1])/(-1 - (-1.1))
```

Out[9]=
        -2.1

The result is just a little smaller than –2.

Here is the slope between (–1.0, f(–1.0)) and (–0.95, f(–0.95)).

```
In[10]:=
          (f[-0.95] - f[-1])/(-0.95 - (-1))
```

Out[10]=
        -1.95

Again, the slope is near –2. This time it is a little larger.

Try adding the line $y = -2(x + 1) + 1$ to the plot of the curve. The line will pass through the point (–1, 1) with a slope of –2.

```
In[11]:=
    Plot[
      {f[x], -2(x + 1) + 1}, {x,-1.1,-0.9},
      Prolog->
        {{PointSize[0.03], Point[ {-1,1} ]}}
      ];
```

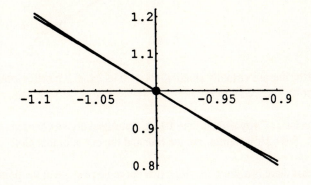

Adding Color to a Plot

Those who have color monitors might like to draw the curve and the line in different colors. The **PlotStyle** option can be set to make the curve red and the line green. **RGBColor** specifies color in proportions of red, green, and blue; its parameters vary from 0 to 1. So **RGBColor[1,0,0]** is pure red and **RGBColor[0,1,0]** is pure green.

```
In[12]:=
    Plot[
      {f[x],-2(x + 1) + 1},{x,-1.1,-0.9},
      PlotStyle->
        {RGBColor[1,0,0], RGBColor[0,1,0]},
      Prolog->
        {{PointSize[0.03], Point[ {-1,1} ]}}
      ];
```

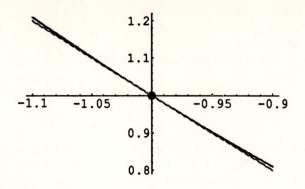

The curve and the line are virtually identical for $-1.05 \leq x \leq -0.95$; in this interval the green line actually covers the red curve.

Practice Exercise 1: Copy and edit the **Plot** command above changing the *x*-interval to $-1.01 \leq x \leq -0.99$. How similar are the line and the curve in this plot?

You see that there is a sense in which this curve has a slope at the point $(-1, 1)$. As we zoom in on curve near this point, magnifying the region in a very small interval, the curve becomes more and more like a line segment with a slope of -2. So, we say that the slope of the curve at $(-1, 1)$ *is* -2.

Let's zoom in on the point $(3, 9)$ and see if we can figure out the slope of the curve at that point.

```
In[13]:=
        Plot[
          f[x],{x,2.99,3.01},
          Prolog->
            {{PointSize[0.03], Point[ {3,9} ]}}
          ];
```

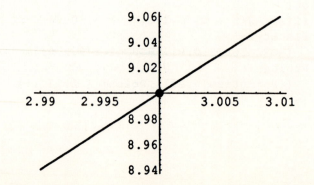

Now, calculate a slopes from (2.99, f(2.99)) to (3, f(3)) and from (3, f(3)) to (3.01, f(3.01)) and see what number these two slopes are near.

```
In[14]:=
        (f[2.99] - f[3.0])/(2.99 - 3.0)
```

```
Out[14]=
        5.99
```

```
In[15]:=
        (f[3.01] - f[3.0])/(3.01 - 3.0)
```

```
Out[15]=
        6.01
```

Sure looks like 6. Check by plotting the line through the point (3, 9) with a slope of 6 along with the curve and see if there is a close fit near the point.

```
In[16]:=
        Plot[
          {f[x],6(x - 3) + 9},{x,2.99,3.01},
          PlotStyle->
            {RGBColor[1,0,0], RGBColor[0,1,0]},
          Prolog->
            {{PointSize[0.03], Point[ {3,9} ]}}
          ];
```

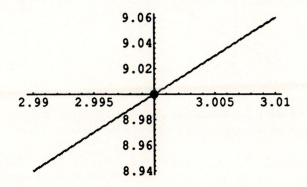

The fit looks perfect. The slope of $y = x^2$ at (3, 9) is 6.

Practice Exercise 2: Find the slope of the curve $y = x^2$ at (2, 4) by the same method used above. (*Ans.* 4)

Tangent Lines to Curves

As you have seen, the line $y = 6(x - 3) + 9$ has a special relationship to the curve $y = x^2$. Both the curve and the line pass through the point $(3, 9)$ and, more importantly, when they are viewed microscopically at this point they are indistinguishable. Mathematicians describe this relationship by saying that the line is *tangent* to the curve at the point. The terminology doesn't make much sense unless we abandon the close-up view and zoom out so that the two plots can be seen over a much wider x-interval.

```
In[17]:=
        Plot[
          {f[x],6(x - 3) + 9},{x,-1,5},
          PlotStyle->
            {RGBColor[1,0,0], RGBColor[0,1,0]},
          Prolog->
            {{PointSize[0.03], Point[ {3,9} ]}}
          ];
```

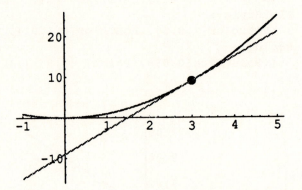

The line just grazes the curve at the point $(3, 9)$ much like the a circle's tangent, which touches it at only one point. In calculus when we speak of the tangent line to a curve at a point, we mean the line through the point which has the same slope as the curve has at the point.

One final example: Find the line tangent to the curve $y = x^3 - 6x^2 + 12x - 3$ at the point $(2, 2)$. Begin by zooming in on the plot so we can figure out the slope of the curve.

```
In[18]:=
        Clear[x,y];
        y[x_]:= x^3 - 6x^2 + 13x - 8;
        Plot[
          y[x], {x, 1.9, 2.1},
          Prolog->
            { {PointSize[0.03], Point[ {2,2} ]} }
          ];
```

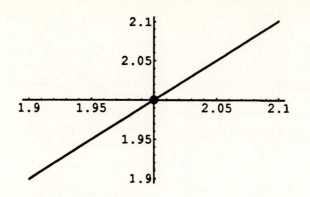

Make a couple of slope calculations.

In[19]:=
```
        (y[2] - y[1.95])/(2 - 1.95)
```

Out[19]=
```
        1.0025
```

In[20]:=
```
        (y[2.01] - y[2])/(2.01 - 2)
```

Out[20]=
```
        1.0001
```

Take the slope of the curve to be 1. The evidence is pretty strong. So, the equation for the tangent line is $y = 1(x - 2) + 2$. Plot the tangent line and the curve.

In[21]:=
```
        Plot[
          {y[x],(x - 2) + 2}, {x,0,4},
          PlotStyle->
            {RGBColor[1,0,0], RGBColor[0,1,0]},
          Prolog->
            { {PointSize[0.03], Point[ {2,2}]} }
          ];
```

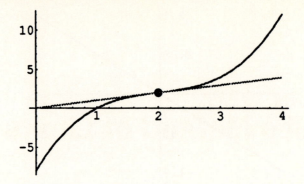

Beautiful!

Code for the Practice Exercises

```
1. Plot[
     {f[x],-2(x + 1) + 1}, {x,-1.01,-0.99},
     PlotStyle->
       { RGBColor[1,0,0], RGBColor[0,1,0] },
     Prolog->
       { {PointSize[0.03], Point[{-1,1}]} }
     ];
```

Problems

1. Estimate the slope of the curve $y = 5x^2$ at the point (1, 5) using a graph with an x-interval of $0.9 \le x \le 1.1$. Using the same x-interval print out a plot of the curve and the tangent line. Then print out a plot of the curve and the tangent line with an x-interval running from -1 to 2 and the point (1, 5) highlighted.

2. Estimate the slope of curve $y = x^3 + 6x + 1$ at the points (0, 1) using a plot with an x-interval of $-0.05 \le x \le 0.05$. Print out two plots showing the curve and its tangent line. One of the plots should zoom in on the point, the other should show the two graphs over a wide x-interval.

3. Print out a plot which shows that the tangent line to the curve $y = x^3$ at the point (2, 8) intersects the curve twice, once at the point of tangency where it just grazes the curve and then at second point where it cuts across the curve.

5

Fermat's Method of Limits

In the previous exploration we saw that when a curve is greatly magnified in a tiny region around a point, it becomes almost straight. By studying these magnified plots we were able to guess the slopes of various curves. Knowing the slope, we could then draw the line tangent to the curve at the point. Our method worked well for the particular curves we studied. But if we are going to make much use of tangent slopes in the future we need a better, more general procedure for finding them, one that can be applied easily and quickly to a variety of functions.

In this exploration we'll take a new look at the problem of tangents and explore an idea that was first set forth by the French mathematician Pierre Fermat. Fermat based his method for finding tangents on the following idea. Suppose a particular curve is given, and suppose *T* denotes the point on the curve where we wish to construct the tangent line. Let *P* denote a nearby point of the same curve and consider the line from *T* to *P*.

Such a line is called a *secant line,* after the name for a line which crosses a circle at two different places. Imagine that the point P moves along the curve toward T while T remains fixed. The secant line TP will rotate closer and closer to the position of the tangent line at T. In the next section we'll explore Fermat's idea further and see exactly how it can be used to determine tangent slopes for curves.

Secant Lines

Consider the problem of finding the line tangent line to $y = x^2$ at the point $(-2, 4)$. We will look at a sequence of secant lines that go from $(-2, 4)$ to another point on the curve (a, a^2), and then see what happens as (a, a^2) moves closer and closer to $(-2, 4)$. The essential idea is that points on the curve near $(-2, 4)$ are almost on the tangent line. So the slope between one of these points and $(-2, 4)$ is almost equal to the slope of the tangent line. And the closer the second point is to $(-2, 4)$, the closer the slope of the secant will be to the slope of the tangent.

Suppose we use $(1, 1)$ as our second point. It is true that $(1, 1)$ is not very close to $(-2, 4)$, but on the other hand it is not very far away either (the point $(100, 10{,}000)$, for example, is a lot farther away), so it is a good place to start. Figure out the equation of the secant line joining $(-2, 4)$ to $(1, 1)$.

First define the function.

```
In[1]:=
        f[x_]:= x^2
```

Then find the slope from $(-2, f(-2))$ to $(1, f(1))$.

```
In[2]:=
        m = (f[1] - f[-2])/(1 - (-2))
Out[2]=
        -1
```

The formula for this secant line is

```
In[3]:=
        secLine[x_]:=-1(x - (-2)) + 4
```

The expression simplifies.

```
In[4]:=
        secLine[x]
Out[4]=
        2 - x
```

Now, plot the curve and the secant line, emphasizing the important points. Give the line and the curve different colors if you wish. (If you do not have a color monitor, try `PlotStyle->{{}, {GrayLevel[.5]}}` instead of the color setting.)

```
In[5]:=
        Plot[
          {f[x], secLine[x]}, {x,-3,2},
          PlotStyle->
            {RGBColor[1,0,0], RGBColor[0,1,0]},
          Prolog->
            {PointSize[.03],
            Point[{-2,f[-2]}],
            Point[{1,f[1]}]
            }
        ];
```

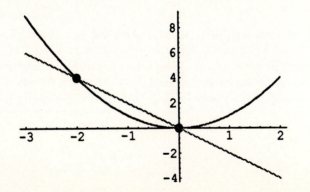

Practice Exercise 1: Copy and edit the commands above to produce a plot of $y = x^2$ and the secant line joining the points $(-2, f(-2))$ and $(0, f(0))$. Emphasize the points $(-2, f(-2))$ and $(0, f(0))$. If you have a color monitor, try changing the parameters for `RGBColor` and see what happens. (*Ans.* The plot should look something like this.)

You have just seen the plots of two secant lines through the point (–2, 4). As we continue our investigation, we are going to look at many secant lines drawn through (–2, 4) in order to see what happens to the line as the second point moves closer and closer to (–2, 4). This will involve finding and plotting all of these lines, so we would like an easy way to refer to an arbitrary secant line. It's not hard to do. Let's repeat the calculations we made in finding the secant line from (–2, 4) to (1, 1), but this time we will make the second point an arbitrary point and call it $(a, f(a))$ instead of $(1, f(1))$. The slope from (–2, 4) to $(a, f(a))$ will, of course, depend upon a.

```
In[6]:=
        Clear[m,a];
        m:= (f[a] - 4)/(a + 2)
```

Once we know what the slope is, we can write the equation for the secant line.

```
In[7]:=
        Clear[secLine,x];
        secLine[x_]:= m(x + 2) + 4
```

The function **secLine[x]** is now a rather complicated expression that depends on a. *Mathematica* will show it to you anytime you want to see it.

```
In[8]:=
        secLine[x]
```

```
Out[8]=
             (-4 + a² ) (2 + x)
        4 + ───────────────────
                  2 + a
```

When a specific value is assigned to a, this expression simplifies to the formula for the secant line. Try $a = 1$.

```
In[9]:=
        secLine[x]/.a->1
```

```
Out[9]=
        2 - x
```

The same line we obtained before! A note about the new construction used here: **/.a->1** causes the **secLine** function to be evaluated for $a = 1$. We'll have a great deal more to say about this method for assigning values to variables in Exploration Six.

Now we can rewrite the **Plot** command so that it will plot the curve and the secant line from (–2, 4) to an arbitrary point $(a, f(a))$. The easiest way to rewrite it is to copy the above **Plot** command and edit it. The only change needed is to replace the 1's in the last line with **a**'s.

Assign a value to **a** before executing **Plot**.

```
In[10]:=
        a = 1;

        Plot[
          {f[x], secLine[x]}, {x,-3,2},
          PlotStyle->
            {RGBColor[1,0,0], RGBColor[0,1,0]},
          Prolog->
            {PointSize[.03],
            Point[ {-2,f[-2]} ],
            Point[ { a,f[ a]} ]
            }
        ];
```

The procedure just developed for plotting a curve with a secant line is summarized below for your convenience. There are three parts: the definitions of **m** and **secLine**, the assignment of values to **a** and **f**, and the **Plot** command.

```
In[11]:=
      Clear[a,m,secLine,x,f];
      m:= (f[a] - 4)/(a + 2);
      secLine[x_]:= m(x + 2) + 4;

      f[x_]:= x^2;
      a = 1;

      Plot[
        {f[x],secLine[x]}, {x,-3,2},
        PlotStyle->
          {RGBColor[1,0,0], RGBColor[0,1,0] },
        Prolog->
          {PointSize[.03],
          Point[ {-2,f[-2]} ],
          Point[ { a,f[ a]} ]
          }
        ];
```

Practice Exercise 2: Copy the above commands into your notebook, execute them, and make sure that your plot agrees with the one shown.

Practice Exercise 3: Copy and edit the commands to give a plot with the secant line drawn from (–2, 4) to (–1, 1). (Note: You only need to change the line that assigns a value to **a**.) Repeat to get the secant line from (–2, 4) to (–1.5, 2.25).

Practice Exercise 4: What happens if you assign $a = -2$ and enter the **Plot** command? Why does it happen?

Packages

We want to generate a large number of plots using our secant plotting procedure so that we can actually watch the secant line rotating into the tangent position. Your *Exploring Calculus* disk contains a package called **SecantLineAn.m** which will be a big help in drawing these plots.

Packages are *Mathematica* files that perform specialized tasks that are not needed .in most *Mathematica* sessions and so are not automatically loaded with the kernel. This means that when you need to use a package, you must ask for it; and if *Mathematica* can not find it, you must assist in the search for it. Finding the package that you need can be very much like searching a file cabinet for a certain letter. You must first find the correct drawer. Once the drawer is open, the proper hanging file folder must be located and then, inside the hanging file folder the manila folder which contains the letter you want to read. With this analogy in mind, you'll have an idea of what is meant when we say that the disk that came with your book has a folder named **Packages**, inside of which there is a package name **SecantLineAn.m**.

Begin the search. The command which loads the package into memory is **<<SecantLineAn.m**. The first two symbols are "less than" signs.

In[12]:=

        <<SecantLineAn.m

A dialog box will appear in which M*athematica* complains that it can't locate the file. Insert the *Exploring Calculus* disk that came with your book. The drive icon in the dialog box will automatically switch to the floppy disk. (If it does not, or if you already had the disk inserted, simply click on the drive button.) You should now see the dialog box below.

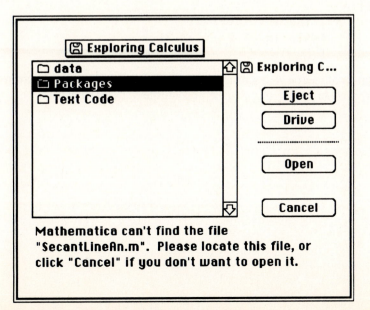

Double click on **Packages**. Then double click on **SecantLineAn.m**. The package should load. The functions defined in the package are now available for use, although the code is not displayed in your notebook. (If you want to see the code, you need to open the package from the file menu.)

Making a Movie

A function called **MultiSecantLinePlot** is in the package just loaded. It will produce a sequence of plots, like the ones we made above. Each plot includes the curve and a secant line. After this sequence of plots is generated, it can be animated to create a moving picture. This function has the form

```
MultiSecantLinePlot[f[x], {x,-3,2}, {-2,2}].
```

The first two arguments **f[x]** and **{x, -3, 2}** are the same as in the ordinary **Plot** command. They give the function to plot and the x-interval for the plot. The third argument **{-2, 2}** specifies two points on the curve: **-2** is the first coordinate of the fixed point (where we want the tangent), and **2** is the first coordinate of the initial position of the moving point. When the command is entered, it will plot the function and a secant line for eight different positions of the moving point.

A note of caution: Producing plots in *Mathematica* takes a lot of memory. It is possible that your computer will run out of memory during this (or later) plotting commands. If that happens, the computer will freeze up, so you will have to restart it and, **as a result, you will lose any work not previously saved. So, if you want to keep the work you have done so far, you should save it before executing the next command.**

```
In[13]:=
        MultiSecantLinePlot[
          f[x], {x,-3,2}, {-2,2}
          ]
```

You should see on your screen the following plots, which are given here in compact form. The labels "a = 1", etc. give the x coordinate of the moving point.

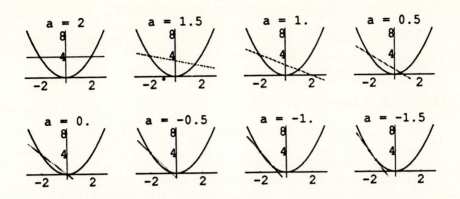

If your system ran out of memory and crashed, restart your machine and reopen your notebook. Free up some memory before you run the command again. The quickest way to do this is to **Cut** all the graphs from your notebook. Leave their input cells so you will be able to redraw them later if you want to.

The **MultiSecantLinePlot** function accepts options from the **Plot** command. In particular, you may include **PlotStyle** options (if you want colored curves, for example) or the **PlotRange** option. You can also select the number of plots to be drawn by placing that number at the end of the point list. For example, if we had used **{-2, 2, 4}** instead of **{-2, 2}**, only four plots would have been drawn. Unless you specify a value, eight plots will be drawn.

ANIMATING A SEQUENCE OF PLOTS

Now that all the plots have been generated, you are ready to animate them. Scroll the screen back to the first plot and click on the bracket that encloses all eight plots; this is the second bracket from the inside. Then pull down the **Graph** menu and release on **Align Selected Graphics**. You will get a dialog box that already has x's in the Top, Left, Right, and Bottom boxes. (If the boxes do not have x's, click on them.) Click on the **OK** button . You are now ready to animate. Pull down the **Graph** menu again and release on **Animate Selected Graphics**.

The animation may be too fast or too slow; or you may want to stop it temporarily. You can control these features using buttons at the bottom left of the window. There are six buttons.

From left to right, their functions are:

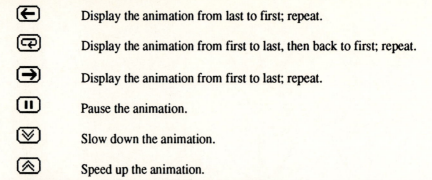

⬅	Display the animation from last to first; repeat.
🔁	Display the animation from first to last, then back to first; repeat.
➡	Display the animation from first to last; repeat.
⏸	Pause the animation.
⌄	Slow down the animation.
⌃	Speed up the animation.

Try clicking on the buttons and observe what happens.

The best way to see the behavior of the secant lines is to control the display by hand. First, pause the display by clicking on the pause button. Then click on the scroll bar at the bottom of the window. By dragging the box to the left or the right in that scroll bar you can select which frame is displayed. Click at various positions in the scroll bar and observe what happens.

Now place the arrow all the way to the left in the bottom scroll bar and click the mouse button. This moves the scroll box to the left end. Then press and hold on the scroll box, and drag it slowly to the right. Observe carefully what happens in the animation. Do you see that the secant line seems to become more and more like the tangent line at $(-2, 4)$ as a gets nearer to -2?

Let's zoom in a closer look, say for $-2.5 \leq x \leq -1.5$. Edit the above **MultiSecantLinePlot** command so that the range is **{x, -2.5, -1.5}**. Unfortunately, the axes of the plot clutter up the very part of the plot we want to see. You can move the axes to a more convenient place by including the option **AxesOrigin->{-2.4, 3}** which will draw the horizontal axis at $y = 3$ and the vertical axis at $x = -2.4$.

```
In[14]:=
      MultiSecantLinePlot[
        f[x], {x,-2.5,-1.5}, {-2,2},
        AxesOrigin->{-2.4,3}
        ]
```

Here is the output in compact form:

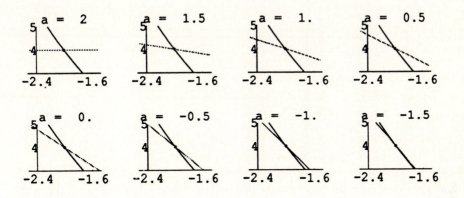

Animate these as before.

Practice Exercise 5: Edit the command to generate a sequence of plots with $-2.5 \le x \le -1.5$ and with the first secant line drawn to $(-1, 1)$. Animate this sequence. In this case the moving point gets closer to $(-2, 4)$. What happens to the secant lines?

Using Limits to Find Tangent Slopes

From these animations it seems clear that the secant lines move closer and closer to the tangent line as a approaches -2. So, how do we use this insight to produce the slope of the tangent line? Why don't we just set $a = -2$ and be done with it? Practice Exercise 4 shows this approach does not work. Here is an idea that will work: Since the secant lines get closer and closer to the tangent line as a approaches -2, the slopes of the secant lines ought also to be getting closer and closer to the tangent slope at $(-2, 4)$. So, construct a table of secant slope values and see what it looks like.

```
In[15]:=
        Clear[a,m];
        m=(f[a]-f[-2])/(a-(-2));
        TableForm[
          Table[{a,m}, {a,-1.0,-1.9,-0.1}],
          TableHeadings->
            {None, {"a\n","secant\nslope\n"} }
          ]
```

Out[15]//TableForm=

```
a          secant
           slope

-1.        -3.
-1.1       -3.1
-1.2       -3.2
-1.3       -3.3
-1.4       -3.4
-1.5       -3.5
-1.6       -3.6
-1.7       -3.7
-1.8       -3.8
-1.9       -3.9
```

It is pretty clear from the pattern in the table entries that as a approaches -2, the secant slope approaches -4.

You can ask *Mathematica* directly for the number which the secant slope is approaching as a nears -2 by using a `Limit` command.

```
In[16]:=
     Clear[a];
     Limit[(f[a]-f[-2])/(a-(-2)),a->-2]
```

Out[16]=
```
     -4
```

One way or the other, Fermat's idea gives -4 for the slope of the curve $y = x^2$ at $x = -2$, $y = 4$. Therefore the tangent line is $y = -4 (x + 2) + 4$.

Define the tangent line function and plot it together with the curve for one final check of our work.

```
In[17]:=
     tanLine[x_]:=-4 (x + 2) + 4
```

```
In[18]:=
     Plot[
       {f[x],tanLine[x]},{x,-3,2},
       PlotStyle->
         {RGBColor[1,0,0], RGBColor[0,1,0]},
       PlotRange->{0, 9},
       Prolog->{PointSize[.03], Point[{-2,4}] }
       ];
```

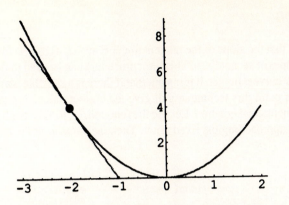

In a very small interval around $x = -2$, the tangent line and the curve are almost indistinguishable, as they should be.

```
In[19]:=
        Plot[
          {f[x], tanLine[x]}, {x,-2.1,-1.9},
          PlotStyle->
            {RGBColor[1,0,0], RGBColor[0,1,0]},
          Prolog->{PointSize[.03], Point[{-2,4}] }
          ];
```

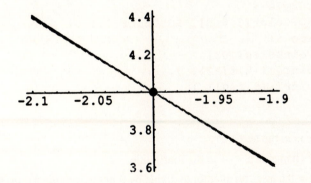

Conclusion

We have seen that the slope of the tangent to $y = x^2$ at $(-2, 4)$ can be found by taking the limit of the slopes of secant lines. This limit procedure can be used to find tangent slopes for many other curves. In fact, it turns out that if there is a sensible way to assign a slope to a curve at a point this technique will give the correct value. The procedure may be written in mathematical notation. Let f be the function, $(x, f(x))$ be the fixed point, and $(a, f(a))$ the point approaching the fixed point. Then the slope m of the tangent line is

```
                f[a]  -  f[x]
         lim[-------------]
                 a  -  x
```

as $a \to x$.

Code for Practice Exercises

```
1. m = (f[-1] - f[-2])/(-1 - (-2));
   secLine[x_]:= m(x - (-2)) + 4;
   Plot[{f[x], secLine[x]}, {x,-3,2},
     PlotStyle->
       {RGBColor[1,0,0], RGBColor[0,1,0] },
     Prolog->
       {PointSize[.03]
       Point[ {-2,f[-2]} ],
       Point[ {0,f[0]} ]
       }
     ];
```

2. The code is in the text.

3. The only change is a = −1 in the fourth line.

4. Setting a = 0 causes an attempt to divide by 0 in the formula for **m**. *Mathematica* gives a variety of interesting error messages.

```
5. MultiSecantLinePlot[
     f[x], {x,-2.5,-1.5}, {-2,-1}
     ]
```

Problems

1. Consider the curve $y = x^2$. Make an animation that shows the secant lines through $(3, 9)$ and $(a, f(a))$, with a starting at 0. Set the x-interval to $-2 \leq x \leq 4$.

2. Look more closely at the point $(3, 9)$ in Problem 1. Make an animation that shows the secant lines through $(3, 9)$ and $(a, f(a))$, with a starting at 2. Set the x-interval to $2 \leq x \leq 4$.

3. Consider the curve $y = x^3 - 6x^2 + 6x + 8$. Make an animation that shows the secant lines through $(2, 4)$ and $(a, f(a))$, with a starting at 5. Choose an appropriate interval for x.

4. Construct a table of values for the slopes of secant lines and use it to find the slope of the tangent line to the curve $y = x^3 - 6x^2 + 6x + 8$ at $x = 2$. Plot the curve and the tangent line.

6

Polynomial Functions and Their Derivatives

In this chapter we'll explore the relationships between graphs of polynomial functions and the graphs of their first derivatives. Begin with the function

$$y = 2x^3 - 21x^2 + 60x + 20, \quad -1 \le x \le 8.$$

Teach *Mathematica* the rule for the function and ask it for a graph.

```
In[1]:=
     y[x_]:= 2x^3 - 21x^2 + 60x + 20;
     Plot[
       y[x], {x, -1, 8},
       AxesLabel->{"x","y"}
       ];
```

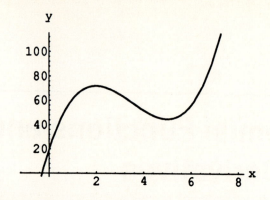

Notice that the graph changes direction twice in this interval. Looking at it from left to right, it rises to a point whose x-coordinate is about 2, changes direction at that point, and falls until x is 5. Then it changes direction again and rises.

At points where the curve is rising, the tangents have positive slope; at points where it is falling, they have negative slope. Here is a graph of the function with its tangent lines displayed at the points where $x = 1$, 3, and 7:

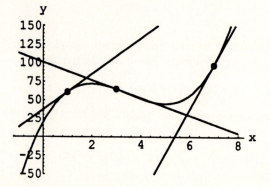

At $x = 1$ the curve is rising, so the tangent line has a positive slope. The curve is falling at the point where $x = 3$ and there you see a tangent line with a negative slope. At $x = 7$ the rising curve again has a positive tangent slope.

The function's derivative will tell us exactly what the slopes of these three lines are. Here's the formula for the derivative:

```
In[3]:=
        y'[x]

Out[3]=
        60 - 42 x + 6 x²
```

The input line below instructs *Mathematica* to evaluate this formula when $x = 1, 3$, and 7.

```
In[4]:=
        {y'[1], y'[3], y'[7]}
```

```
Out[4]=
        {24, -12, 60}
```

These are the exact slopes of the three tangent lines: positive 24, negative 12, and positive 60.

Plotting Functions and Their Derivatives

The **Plot** commands below graphs the function

$$y = 2x^3 - 21x^2 + 60x + 20$$

and its derivative

$$y' = 6x^2 - 42x + 60$$

over the same x-interval, $-1 \le x \le 8$.

```
In[5]:=
        Plot[
          y[x], {x, -1, 8},
          AxesLabel->{"x","function"}
          ];
        Plot[
          y'[x], {x, -1, 8},
          AxesLabel->{"x","derivative"}
          ];
```

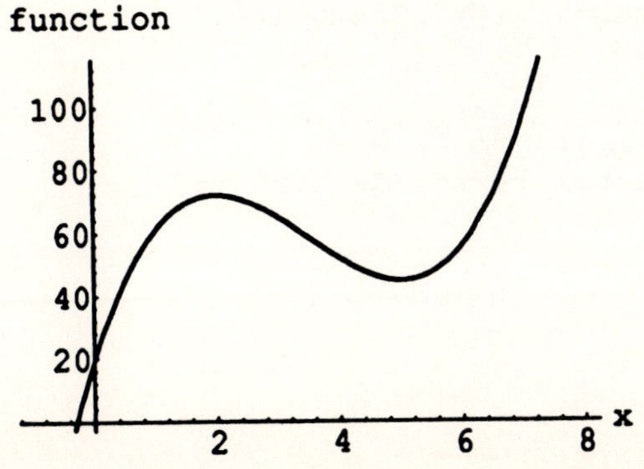

derivative

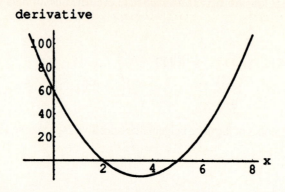

Study the curves carefully. Look at the graph of y': It lies above the x-axis for $-1 \le x \le 2$ and again when $5 \le x \le 8$ indicating that the derivative is positive for these values of x. Notice how the graph of y behaves on the same intervals; it is rising. The graph of y falls for $2 \le x \le 5$, exactly the interval over which y' is negative. The derivative's x-intercepts, the numbers at which the graph y' crosses the x-axis, correspond to the points on the graph of y where it changes direction.

Let's try another example.

Graph the function

$$y = x^5 + 2x^4 - 6x^3 + 2x - 5$$

and its derivative

$$y' = 5x^4 + 8x^3 - 18x^2 + 2$$

Over the interval $-1.5 \le x \le 3$.

```
In[6]:=
    Clear[x,y];
    y[x_] := x^5 + 2x^4 - 6x^3 + 2x - 5;
    Plot[
      y[x], {x, -1.5, 3},
      AxesLabel->{"x","function"}
      ];
    Plot[
      y'[x], {x, -1.5, 3},
      AxesLabel->{"x","derivative"}
      ];
```

function

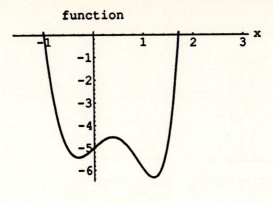

derivative

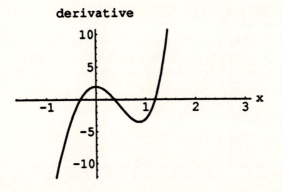

Look at the plot for y. It falls, rises for a bit, falls again, and then rises as x increases in value. Correspondingly, the values of the derivative are negative, positive, negative, and positive. The points were the graph of the function changes direction correspond to points where the derivative crosses the x-axis.

Here is a third example which illustrates the fact that a curve will not always change direction where its slope is zero.

```
In[7]:=
      Clear[x, y];
      y[x_] := (x - 2)^3 + 1;
      Plot[
        y[x], {x, -.5, 3},
        AxesLabel->{"x","function"}
        ];
      Plot[
        y'[x], {x, -.5, 3},
        AxesLabel->{"x","derivative"}
        ];
```

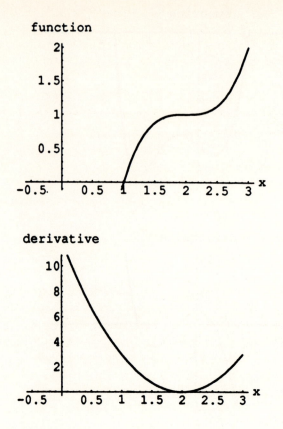

Notice that the curve *y* does not change direction (it is always increasing) even though the derivative is zero when *x* is 2.

In[8]:=

    y'[2]

Out[8]=

    0

    The curve rises to a point above 2 where it levels off momentarily, but it does not fall from this point; it continues to rise. This occurs because the derivative is a positive number for *every* value of *x* other than 2. You can see that the graph of the derivative lies above the *x*-axis except where it just touches it at 2. This behavior is dictated by the derivative's formula.

In[9]:=

    y'[x]

Out[9]=

    $3\ (-2 + x)^2$

Every derivative value is 3 times a perfect square and so, as long as x is different from 2, it will always have a positive value.

We will move on now to analyzing polynomial functions of higher degree. But first take a minute to study a new and very useful *Mathematica* command.

Mathematica's Replacement Command

Observe the following input-output sequence.

```
In[10]:=
        z = t^2;
        z/.t->3

Out[10]=
        9
```

The new symbol used /. is a two-keystroke symbol: / (next to the right-hand shift key) followed by a period. The first line of input defines z to be t^2 and the second asks for z's value when t is 3. *Mathematica* obligingly returns the correct number, 9. Simple enough, and now you have an idea of what the strange-looking symbol /. means. It says: Evaluate the expression on the left of the symbol /. according to the assignment on the right.

Here's another use of /. In this case it returns a list of z-values.

```
In[11]:=
        z/.{{t->3},{t->-1},{t->0}}

Out[11]=
        {9, 1, 0}
```

Let's get a little fancier.

```
In[12]:=
        list1 = {{t->3},{t->-1},{t->0}};
        {t,z}/.list1

Out[12]=
        {{3, 9}, {-1, 1}, {0, 0}}
```

The output is the original list, **list1**, of t values paired with their corresponding z values.

Before going on, **Clear** the variables used here.

```
In[13]:=
        Clear[z,t,list1]
```

How to Use the Derivative to Sketch Curves

Consider the following problem: Find out where the function

$$y = 5.0x^6 - 24x^5 - 165x^4 + 1120x^3 - 765x^2 - 5400x + 100$$

is rising, where it is falling, and at what points it changes direction. Graph the function and its derivative.

Teach *Mathematica* the rule for the function.

```
In[14]:=
        Clear[x,y];
        y[x_]:= 5.0x^6 - 24x^5 - 165x^4 +
                1120x^3 - 765x^2 - 5400x + 3000
```

Don't leave out the decimal point in `5.0x^6`. It will insure that *Mathematica* returns decimal solutions to any algebra problems involving this function and will save you the trouble of having to use **N** with **Solve** in order to avoid those unreadable exact solutions.

Find the points on the curve where the tangent slope is zero. First make a list, call it **list2**, of the values of x for which the derivative is 0.

```
In[15]:=
        list2 = Solve[y'[x]==0, x]
```

```
Out[15]=
        {{x -> -5.}, {x -> -1.}, {x -> 3.},
         {x -> 3.}, {x -> 4.}}
```

Now, make a list of the points on the curve which have these numbers as their first coordinates.

```
In[16]:=
        list3 = {x, y[x]}/.list2
```

```
Out[16]=
        {{-5., -79125.}, {-1., 6379.}, {3., -5397.},
         {3., -5397.}, {4., -5496.}}
```

At these five points the tangent line has a zero slope. Wait a minute, count again. Two of them are the same point, so there are actually only four different points.

Execute a **Plot** command that is sure to show all of these points. There are lots of possibilities, but you can see that the x-range must include the interval $-5 \le x \le 4$. Try this one:

```
In[17]:=
        Plot[
          y[x], {x,-6,5},
          AxesLabel->{"x","y(x)"}
          ];
```

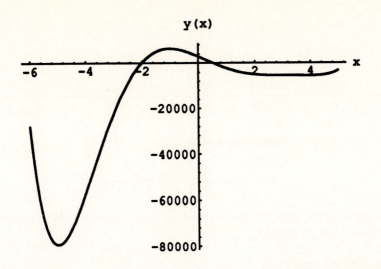

The direction changes are clearly visible at $x = -5$ and $x = -1$. The behavior of the plot at $x = 3$ and $x = 4$ is more difficult to see. The problem is that vertical scale is too large to show the variations in the curve at these points. Look again at **list3**.

```
In[18]:=
        TableForm[list3]
```

```
Out[18]=
        -5.     -79125.
        -1.     6379.
        3.      -5397.
        3.      -5397.
        4.      -5496.
```

The highest and lowest y-values differ by about 85,000. The y-values at $x = 3$ and 4 differ only by about 100 or so. If you zoom in on the interval $3 \le x \le 4$, the details of the curve become clear.

```
In[19]:=
        Plot[
          y[x], {x,2.5,4.5},
          AxesLabel->{"x","y(x)"}
          ];
```

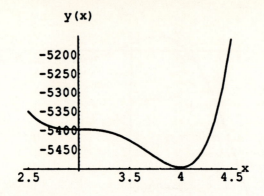

So, there it is—a direction change when x = 4 but no change at x = 3. (Notice that the axes do not intersect at the origin.)

Look at the graph of the derivative over the same interval to see why there is no direction change when x = 3.

In[20]:=
```
        Plot[
          y'[x], {x, 2.5, 4.5},
          AxesLabel->{"x", "y'"}
          ];
```

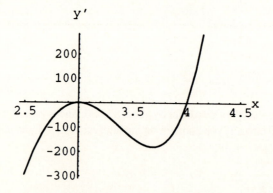

The derivative changes sign at 4 but not at 3.

In the assignment you'll be asked to put these ideas to work analyzing and writing about the behavior of polynomial curves.

Changing an Input Cell to a Text Cell

When you are writing ordinary text in *Mathematica*, in other words when you are using *Mathematica* as a word processor, it is usually more convenient to change the cell you are typing in to what is called a **Text** cell. This section explains how this can be done.

As you begin typing in a new cell, *Mathematica* assumes that you are typing a command which you will eventually ask it to execute, so it automatically gives the cell the characteristics of an **Input** cell. These characteristics include such things as the margin widths for the cell, the type font used, and its size and face and so forth. You can change the cell type to **Text** using the **Cell Style** menu which is under **Styles** in the main menu bar. Here is how it is done: Select the cell you wish to change by clicking on its cell bracket, pull down the **Style** menu from the main menu bar, drag to the right on **Cell Style** and then down on this menu. Release on the word **Text**. The cell which you previously selected will now be a **Text** cell. Try it.

If you have started typing in a cell, and then decide that you wish it to be a **Text** cell, all you have to do is hold down the Option key, ⌘-key, and 7-key simultaneously.

When you create a new cell, you can specify that it should be a **Text** cell. Move the cursor down the screen until it becomes a horizontal bar. Before you begin typing hold down the ⌘-key and press 7. Now when you begin typing in the new cell it will be a **Text** cell.

You might like to experiment with cells styles. You can see that there are a number of them in on the **Cell Styles** menu.

Problems

1. For each of the following polynomial functions, print out a graph of the function and a graph of its derivative, and write a paragraph explaining the connection between them.

    (a) $y = -6 + 11x - 6x^2 + x^3$

    (b) $y = 8 - 4x - 10.0x^2 + x^3 + 4.0x^4 + x^5$

2. Analyze the polynomial function

$$y = -157500.0x + 99750.0x^2 - 22925x^3 - 1785x^4$$
$$+ 1722x5 - 280x^6 + 15x^7$$

    Graph it and its derivative in as many parts as is necessary to clearly show the behavior of the function and its derivative. Write a paragraph explaining the connection between the two curves.

3. Below are plots of two functions. For each make a rough hand sketch of its derivative.

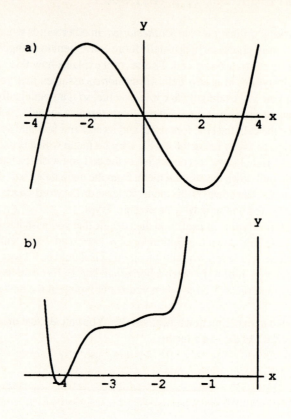

7

Rational Functions and Asymptotes

A *rational function* is one whose rule is the quotient of two polynomials. Such functions have some special and quite dramatic characteristics which we will explore in this section. Begin with the function $f(x) = 3x/(x-2)$. Notice that it is undefined when $x = 2$ because this value of x makes the denominator equal to zero. Ask *Mathematica* to evaluate the function at 2 and see what happens.

```
In[1]:=
        Clear[f,x];
        f[x_] := 3x/(x-2);
        f[2]

Power::infy:
                              1
        Infinite expression - encountered.
                              0

Out[1]=
        ComplexInfinity
```

Before the output is shown, *Mathematica* warns you that you are asking it to divide by 0. The output itself, `ComplexInfinity`, tells you that for values of x near 2 the corresponding function value will be very large. Graph the function for an interval including 2 to see the rapid increase in absolute function values.

```
In[2]:=
      Plot[
        f[x], {x,1,3},
        AxesLabel->{"x","f(x) = 3x/(x-2)"}
        ];
```

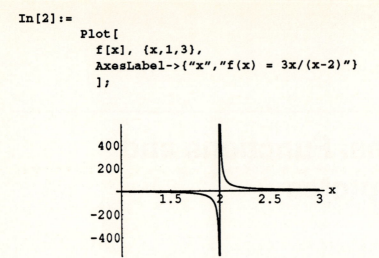

The first thing to note is that what appears to be a vertical line at $x = 2$ is *not* part of the graph of the function. It can't be; the function is undefined at 2. So, you ask, if it isn't part of the graph why does *Mathematica* draw it in? Good question, and the honest answer is that nothing is perfect, even *Mathematica*'s **Plot** command. The command is programmed in such a way that the line gets drawn even though it shouldn't be there. It is a matter of trade-offs. In order to get some of the truly wonderful features that the **Plot** command does have, this is one of the little defects we have to live with. That's life.

All that having been said, the fact is that the vertical line can actually be quite useful in characterizing the function's behavior near $x = 2$ even though it is not part of the graph. To see exactly how this works, you need to look further at the behavior of the function near 2. As is always the case in studying functions, *Mathematica*'s **Table**, **Plot**, and **Limit** commands are a great help.

The first table shows functions values for values of x slightly larger than 2.

```
In[3]:=
      Clear[x,k];
      x = 2 + 10.^-k;
      TableForm[
        Table[{x,f[x]}, {k,1,5}],
        TableHeadings->{None, {"x\n","f[x]\n"}}
        ]
```

```
Out[3]//TableForm=
          x            f[x]

          2.1          63.
          2.01         603.
          2.001        6003.
          2.0001       60003.
          2.00001      600003.
```

The closer x is to 2, the larger the corresponding function value.

Now make a table for x values that are slightly less than 2. **Copy** and **Paste** the previous **Table** command and edit it to produce the command below. (You need make only one small change: the + in the second line becomes a - .)

```
In[4]:=
        Clear[x,k];
        x = 2 - 10.^-k;
        TableForm[
          Table[{x, f[x]}, {k,1,5}],
          TableHeadings->{None, {"x\n","f[x]\n"}}
          ]

Out[4]//TableForm=
          x            f[x]

          1.9          -57.
          1.99         -597.
          1.999        -5997.
          1.9999       -59997.
          1.99999      -599997.
```

As x nears 2 from below, the corresponding functions values are negative, but they become larger and larger in absolute value. You could imagine continuing either one of these tables indefinitely. The pattern in the entries is clear: As x nears 2, the absolute values of $f(x)$ becomes infinitely large. They shoot up to positive infinity for x's on the right side of 2 and down to negative infinity for x's on the left side of 2.

Let's see what *Mathematica*'s **Limit** says about this function. Ask it to tell you what number the values of $f(x)$ are approaching as x nears 2 through numbers which are *smaller* than 2 by including the option **Direction->1**.

```
In[5]:=
        Clear[x];
        Limit[f[x], x->2, Direction->1]

Out[5]=
        -Infinity
```

To get the limit of $f(x)$ as x nears 2 from above, copy the cell above and change the direction to -1

```
In[6]:=
        Clear[x];
        Limit[f[x], x->2, Direction->-1]
```

```
Out[6]=
        Infinity
```

The output from the **Plot**, **Table**, and **Limit** commands all tell the same tale: Near $x = 2$ the curve is very nearly vertical. For this reason, the vertical line $x = 2$ approximates the curve near 2. Such approximating lines are called *vertical asymptotes*.

So far in thinking about this rational function we have concentrated on its behavior at $x = 2$, since we were aware from the beginning that something unusual was happening there. Let's change our point of view. Zoom out and take a more global look at the function.

```
In[7]:=
        Plot[
          f[x], {x,-100,100},
          AxesLabel->{"x","f[x] = 3x/(x-2)"}
          ];
```

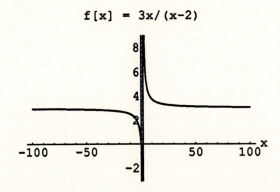

Notice that the vertical line at $x = 2$ is quite close to the vertical axis. This is caused by the relatively large scale on the horizontal axis. Zoom out again.

```
In[8]:=
        Plot[
          f[x], {x,-10000,10000},
          AxesLabel->{"x","f[x] = 3x/(x-2)"}
          ];
```

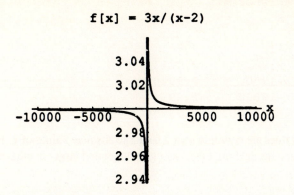

f[x] = 3x/(x-2)

For very large and very small values of *x*, the corresponding function values seem to be very close to 3. Check with a **Table**.

```
In[9]:=
      Clear[x,k];
      x = 10.^k;
      TableForm[
        Table[{x, f[x]}, {k,1,5}],
        TableHeadings->{None, {"x\n","f[x]\n"}}
        ]
```

```
Out[9]//TableForm=
           x            f[x]

           10.          3.75
           100.         3.06122
           1000.        3.00601
           10000.       3.0006
           100000.      3.00006
```

```
In[10]:=
      Clear[x,k];
      x = -10.^k;
      TableForm[
        Table[{x, f[x]}, {k,1,5}],
        TableHeadings->{None, {"x\n","f[x]\n"}}
        ]
```

```
Out[10]//TableForm=
            x               f[x]

            -10.            2.5
            -100.           2.94118
            -1000.          2.99401
            -10000.         2.9994
            -100000.        2.99994
```

Function values are certainly near 3 for large absolute values of x. Try another test: Ask *Mathematica* for the limit of $f(x)$ as x gets larger and larger in the positive direction.

```
In[11]:=
        Clear[x];
        Limit[f[x], x->+Infinity]

Out[11]=
        3
```

How about when x gets more and more negative?

```
In[12]:=
        Clear[x];
        Limit[f[x], x->-Infinity]

Out[12]=
        3
```

Clearly, as the values of x get large in absolute value, the values of $f(x)$ approach 3. This means that the line $y = 3$ approximates the curve for very large absolute values of x. This horizontal line is, therefore, also an asymptote for the function. It is referred to as a *horizontal asymptote*.

Recapping, we have seen that the function $f(x) = 3x/(x-2)$ has a vertical asymptote $x = 2$ and a horizontal asymptote $y = 3$. Here is a **Plot** command which shows this asymptotic behavior:

```
In[13]:=
        Plot[
          {f[x],3}, {x,-5,8},
          PlotStyle->{{}, {Dashing[{.03}]}},
          AxesLabel->{"x", "f[x] = 3x/(x-2)"},
          PlotRange->{-15,25}
          ];
```

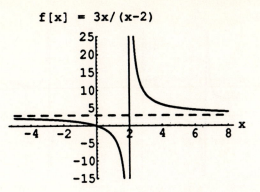

f[x] = 3x/(x-2)

The **PlotStyle** setting forces the horizontal asymptote to show as a dashed line.

Not All Asymptotes Are Lines

Consider another rational function, $g(x) = (x^4 - x^3 - 6x^2 + 1)/(x^2 - x - 6)$.

```
In[14]:=
        Clear[g,x];
        g[x_]:= (x^4 - x^3 - 6x^2 + 12)/
                (x^2 - x - 6)
```

Vertical asymptotes are likely to occur at the values of x which make the denominator equals to zero. Find these values for $g(x)$.

```
In[15]:=
        Solve[Denominator[g[x]]==0, x]
```

```
Out[15]=
        {{x -> -2}, {x -> 3}}
```

Look at a plot over an interval containing these values of x.

```
In[16]:=
        Plot[
          g[x], {x,-3,4},
          AxesLabel->{"x","g(x)"}
          ];
```

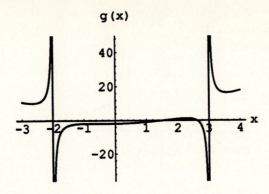

Not bad for a rough sketch. The expected asymptotes are there. Now get a global view of the function.

```
In[17]:=
        Plot[
          g[x], {x,-100,100},
          AxesLabel->{"x","g(x)"}
          ];
```

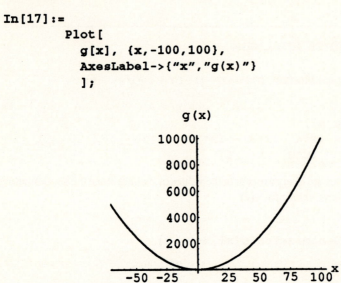

For large values of x, the function $g(x)$ behaves like the curve above. Let's find out what curve it is. Look at the rule for $g(x)$.

```
In[18]:=
        g[x]
```

```
Out[18]=
```
$$\frac{12 - 6 x^2 - x^3 + x^4}{-6 - x + x^2}$$

Notice that the polynomial in the numerator is of degree 4, larger than that of the polynomial in the denominator whose degree is 2. This means that you can divide the bottom polynomial into the top one to obtain a polynomial quotient and remainder. Here are the *Mathematica* commands for getting the quotient, which is assigned the variable **q**, and the remainder, which is assigned the variable **r**.

```
In[19]:=
        q = PolynomialQuotient[
            Numerator[g[x]],
            Denominator[g[x]], x]

Out[19]=
        x²
```

```
In[20]:=
        r = PolynomialRemainder[
            Numerator[g[x]],
            Denominator[g[x]], x]

Out[20]=
        12
```

Now, rewrite the rule for $g(x)$ as its quotient plus its remainder divided by its denominator.

```
In[21]:=
        q + r/Denominator[g[x]]

Out[21]=
                    12
        x²  + ─────────────
              -6 - x + x²
```

That is what $g(x)$ looks like divided out. Check by doing the addition to make sure that you get $g(x)$ back.

```
In[22]:=
        Together[%]

Out[22]=
        12 - 6 x²  - x³  + x⁴
        ─────────────────────
             -6 - x + x²
```

Good. It checks. That is $g(x)$ in its original form. So, we know that the sum

$$x^2 + 12/(x^2 - x - 6)$$

is another way to write the rule for $g(x)$.

The advantage of dividing out the expression for $g(x)$ in this fashion is that when it is written in this form it is much easier to tell what is happening to function values for large values of x. Imagine substituting a large (in absolute value) number for x into the expression

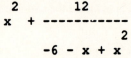

$$x^2 + \frac{12}{-6 - x + x^2}$$

The value returned will be x^2 with a very small fraction tacked on at the end, which means that the value of the sum will be very close to x^2. Thus, for large enough values of x, the value of $g(x)$ is approximated by x^2. So, $y = x^2$ is an asymptote for the function.

Here is a plot which shows these features of the curve:

```
In[23]:=
        Plot[
            {g[x],x^2}, {x,-10,10},
            PlotStyle->{{},{Dashing[{.03}]}},
            AxesLabel->{"x","g[x]"},
            PlotRange->{-40,60}
            ];
```

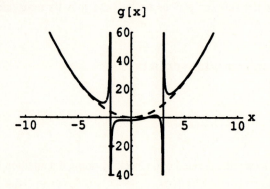

Notice that for large absolute values of x, the curve for $g(x)$ and the dashed curve for x^2 are indistinguishable.

There Won't Always Be Vertical Asymptotes

Example 1: The function $h(x) = 1/(x^2 + 1)$ has no vertical asymptotes.

```
In[24]:=
        Clear[h,x]
        h[x_]:= 1/(x^2 + 1)
        Solve[Denominator[h[x]]==0, x]
```

```
Out[24]=
        {{x -> I}, {x -> -I}}
```

The solutions are not real. So, there are no numbers on the *x*-axis which will make the denominator equal to zero. Here is a plot of the curve:

```
In[25]:=
        Plot[h[x], {x, -3,3},
            AxesLabel->{"x","h(x)=1/(x^2+1)"}];
```

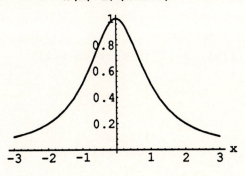

h(x)=1/(x^2+1)

Example 2: Here is a function which has no vertical asymptotes even though the function is undefined at $x = 2$:

```
In[26]:=
        Clear[F,x];
        F[x_]:= (2 + x - x^2)/(6 - 3x);
        F[2]
```

```
Power::infy:
                                    1
            Infinite expression - encountered.
                                    0
```

```
Out[26]=
        Indeterminate
```

However, the graph near 2 shows no sign of a vertical asymptote.

```
In[27]:=
        Plot[F[x], {x,1,3}];
```

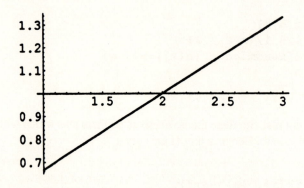

The absence of a vertical asymptote is not a mistake on *Mathematica*'s part. Check the limiting values of $F(x)$ as x nears 2, and you will see that the function values do not shoot to infinity.

```
In[28]:=
        Limit[F[x], x->2, Direction->-1]
```

```
Out[28]=
        1
```

```
In[29]:=
        Limit[F[x], x->2, Direction->1]
```

```
Out[29]=
        1
```

As x approaches 2, the function values are approaching 1, not infinity. So there can't be a vertical asymptote at 2.

Look at the graph again. The function $f(x)$ is undefined at 2, so the point $(2,1)$ is not on this plot. *Mathematica* cannot show a single missing point, so you must remove it in your mind's eye. The plot really looks like this:

```
In[30]:=
        Plot[
          F[x], {x,1,3},
          PlotRange->{0,2},
          Prolog->{PointSize[.03], Point[{2,1}]},
          AxesLabel->{"x","f(x)"}
          ];
```

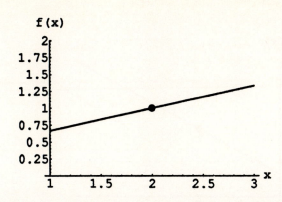

It is understood that the highlighted point is missing from the graph.

In describing this kind of behavior, mathematicians say that the function $F(x)$ has a *point discontinuity* at 2. This means that the break in the curve is produced by a single missing point. When a vertical asymptote occurs, the break is much more dramatic. Such a discontinuity is called an *essential discontinuity*.

What is it in the rule for $F(x)$ which makes the discontinuity at 2 such a mild one? The clue is in the factors of the numerator and denominator.

```
In[31]:=
        Factor[Numerator[F[x]]]

Out[31]=
        (2 - x) (1 + x)


In[32]:=
        Factor[Denominator[F[x]]]

Out[32]=
        3 (2 - x)
```

The common factor $(2 - x)$ can be cancelled out, for all values of x except 2. So, for every other value of x, the rule for $F(x)$ reduces to $(1 + x)/3$. Plot both $F(x)$ and $(1 + x)/3$, and you will see that the difference occurs only at 2.

```
In[33]:=
        Plot[
          F[x], {x,0,5},
          PlotRange->{0,2},
          Prolog->{PointSize[.03], Point[{2,1}]},
          AxesLabel->{"x","f(x)"}
          ];
        Plot[
          (1 + x)/3, {x,0,5},
          PlotRange->{0,2},
          AxesLabel->{"x","(1+x)/3"}
          ];
```

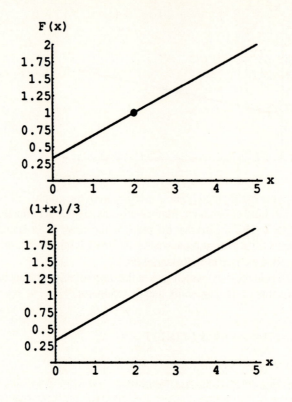

How to Divide and Merge Cells

In the next set of problems you will be asked to study certain rational functions and to write up the results of your investigations illustrating your remarks with tables, plots, and limits. Suppose that the following circumstance occurs: After writing several sentences, you realize that you would like to insert a graph between your second and third sentences. Before the insertion of the graph your work looks like this:

> This is sentence one. This is sentence two. This is sentence three. And following
> sentence three there are several more sentences

You can make room for the graph by dividing the existing cell into two new cells. Place the cursor between sentences two and three and click. Now, pull down the **Cell** menu and release on **Divide Cell**. It will take a second, but the cell will split apart.

This is sentence one. This is sentence two.

This is sentence three. And following sentence three there are several more sentences

Position the cursor between the two cells and click when it flips to the horizontal I-beam. Type in the command for the graph you wish to insert and press *enter*.

This is sentence one. This is sentence two.

    Plot[.....|

This is sentence three. And following sentence three there are several more sentences

Cells can also be merged. Suppose you wish the following four sentences to be contained together in the same text cell.

This is sentence one.

This is sentence two.

This is sentence three.

This is sentence four.|

Select the four cells. Place the pointer in the top cell bracket. Press and hold the mouse button while you drag down though the lower three brackets. Then release the mouse button. All four cell brackets should be highlighted. Pull down the **Cell** menu and release on **Merge Cells**.

This is sentence one.
This is sentence two.
This is sentence three.
This is sentence four.

Problems

1. Investigate the asymptotic behavior of the rational function $f(x) = 5/(x + 1)$. Use the **Table**, **Plot**, and **Limit** commands in your analysis. Write a paragraph describing your conclusions, and print it along with graphs which illustrate your remarks.

This is the expression for the velocity for values of *t* less than 60. The next calculation gives the velocity for the larger values of *t*.

```
In[11]:=
        h2'[t]
```

```
Out[11]=
        4800 - 32 (-60 + t)
```

Obviously, that expression can be simplified.

```
In[12]:=
        Simplify[%]
```

```
Out[12]=
        6720 - 32 t
```

To the left of $t = 60$ the velocity curve is the line $80t$, but to the right of $t = 60$ it is a different line, $6720 - 32t$. Hence an abrupt change occurs in the direction of the curve.

Question 4: What is the rocket's maximum velocity?

From the graph it is obvious that the maximum velocity occurs when $t = 60$. The actual numerical value, 4800 ft/sec, can be computed from either part of the velocity formula.

```
In[13]:=
        h1'[60]
```

```
Out[13]=
        4800.
```

```
In[14]:=
        h2'[60]
```

```
Out[14]=
        4800
```

Question 5: What is the rocket's acceleration?

Acceleration is the instantaneous rate of change of velocity and is computed by differentiating the velocity formula. In this case acceleration will have a two-part rule just as velocity does. You can see the two parts explicitly by asking *Mathematica* to compute the derivative of velocity.

```
In[15]:=
        h1''[t]
```

```
Out[15]=
        80.
```

```
In[16]:=
        h2''[t]
```

```
Out[16]=
        -32
```

The first output tells you that for $t < 60$, the velocity increases at the rate of 80 ft/sec for each second the rocket is aloft. The rocket engines are pushing it to go faster and faster. After 60 seconds, velocity decreases at the rate of 32 ft/sec. Gravity decelerates the rocket. Here is a plot of the acceleration:

```
In[17]:=
        p5 = Plot[
                h1''[t], {t,0,60},
                DisplayFunction->Identity
                ]
        p6 = Plot[
                h2''[t], {t,60, 387},
                DisplayFunction->Identity
                ]
        accel = Show[
                {p5,p6},
                DisplayFunction->$DisplayFunction,
                AxesLabel->{"t","acceleration"}
                ];
```

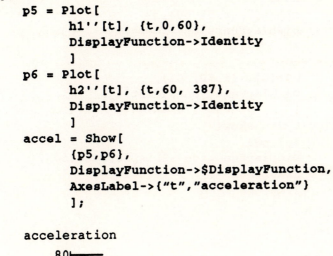

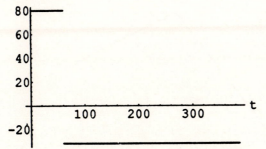

The very abrupt change in acceleration is clear at $t = 60$ seconds.

Question 6: What is the acceleration when *t* = 60 ?

It is impossible to tell from the acceleration curve itself because of the discontinuity at 60. Coming in from the right it looks like the acceleration should be –32 ft/sec², but viewed from the left, it clearly should be 80 ft/sec².

Maybe the velocity curve will be more help. The acceleration at 60 is numerically the same as the slope of the line tangent to the velocity curve at 60. The tangent line, in turn, is the line which the curve approaches at we zoom in on the point (60, 4800). Let's try zooming in on this point. Maybe that will give us a clue to the nature of the tangent slope at this point. Scroll back to the cell which drew the first velocity curve , **Copy** and **Paste** it into your notebook, and reset the t-ranges for a narrow interval around 60. You had better rename the plots because this curve will be different from the previous ones.

```
In[19]:=
        p7 = Plot[
                h1'[t], {t,59,60},
                DisplayFunction->Identity
                ]
        p8 = Plot[
                h2'[t], {t,60, 61},
                DisplayFunction->Identity
                ]
        velocity2 = Show[
                {p7,p8},
                DisplayFunction->$DisplayFunction,
                AxesLabel->{"t","velocity"}
                ];
```

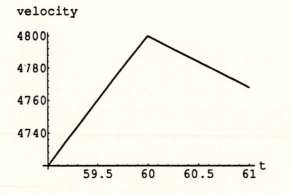

Zoom in a little closer. Narrow the interval again.

```
In[20]:=
      p9 = Plot[h1'[t], {t,59.9,60},
            DisplayFunction->Identity]
      p10 = Plot[h2'[t], {t,60, 60.1},
            DisplayFunction->Identity]
      velocity2 = Show[{p9,p10},
            DisplayFunction->$DisplayFunction,
            AxesLabel->{"t","velocity"}
                ];
```

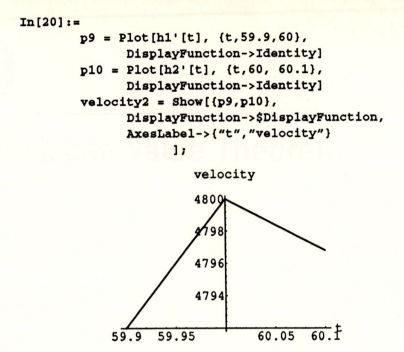

You can try zooming in a couple more times, but the problem is obvious already. The velocity curve does not straighten out as we zoom in on the point (60, 4800). Therefore, it has no tangent line there and so its derivative is undefined at that point. This means the acceleration is undefined when $t = 60$.

The point (60, 4800) is called a *singularity* for the acceleration curve, meaning a point where no acceleration value can be computed.

Problems

1. After t minutes of flight, a certain rocket's height $h(t)$, measured in miles, is given by the two-part rule:

$$h(t) = -t^3/10 + 18t \qquad\qquad 0 \leq t \leq 10$$
$$h(t) = 80 - 12(t - 10) - 16(t - 10)^2 \qquad t > 10$$

   (a) Graph the rocket's height, velocity, and acceleration formulas.
   (b) Write a paragraph analyzing the motion of the rocket in which you discuss all of the important features of the three graphs and their interrelationships.

2. When the faucet is turned on, water runs into the bathtub at the rate of 3 gallons per minute. After 15 minutes the plug is pulled, and water begins draining out at the rate of 5 gallons per minute. Assuming that the faucet remains on during this time, find the rule for the function $V(t)$ which gives the volume of water in the tub after t minutes. Plot graphs of the function and its derivative, and write a paragraph explaining what the plots say about the volume of water as a function of time.

(distance)/(time) = $(f(b) - f(a))/(b - a)$. Then our claim is that for at least one time $t = c$ the instantaneous velocity $f'(c)$ is equal to $(f(b) - f(a))/(b - a)$.

Look at an example.

Example: Suppose an object is moving on a straight line path so that at time t seconds its distance s in feet from the origin is given by the formula $s(t) = t^3$, for $0 \leq t \leq 10$.

Teach *Mathematica* the distance formula.

```
In[1]:=
        s[t_]:= t^3
```

In the 10-second time interval the object moves 1000 feet.

```
In[2]:=
        s[10]
```

```
Out[2]=
        1000
```

Here is the average velocity during the first 10 seconds:

```
In[3]:=
        (s[10] - s[0])/(10 - 0)
```

```
Out[3]=
        100
```

The object moves over the interval with an average speed of 100 feet per second. Keep in mind that its instantaneous speed changes constantly. Define the velocity function and then look at a table of instantaneous velocity values.

```
In[4]:=
        v[t_]:= s'[t]
```

```
In[5]:=
        TableForm[
          Table[{t,v[t]}, {t,0,10}],
          TableHeadings->{None, {"t\n","v(t)\n"}}
          ]
```

```
Out[5]//TableForm=
          t      v(t)

          0      0
          1      3
          2      12
          3      27
          4      48
          5      75
          6      108
          7      147
          8      192
          9      243
          10     300
```

The object accelerates dramatically. After 1 second it is moving at only 3 feet per second, a relative snail's pace; but at 10 seconds, its instantaneous velocity is a very speedy 300 feet per second.

At some time between 5 and 6 seconds the instantaneous velocity is exactly 100 feet per second, the average velocity over the 10-second interval. You can find this time by giving the command:

$$N[Solve[v[t]==100, t], 10]$$

A Geometric Interpretation of Average and Instantaneous Velocity

We will now look more carefully at the relationship between average velocity and instaneous velocity. It is helpful to plot the distance function and give a geometric interpretation to both the average velocity and the instantaneous velocity.

```
In[6]:=
      Plot[
        s[t], {t,0,10},
        AxesLabel->{"t", "s(t)"},
        PlotLabel->"s(t) = t^3",
        Epilog->
          {PointSize[.03],
          Point[{0,0}],
          Point[{10,1000}]
          }
      ];
```

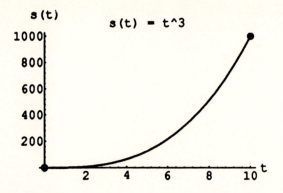

The slope of the line joining these two points is numerically equal to the moving object's average velocity: slope $= m = (s(10) - s(0))/(10-0) = 100 =$ average velocity. The equation of this secant line is $y = m(t - 0) + s(0)$. Define the line as follows:

```
In[7]:=
    m = (s[10] - s[0])/(10 - 0);
    secLine[t_]:= m (t - 0) + s[0]
```

The next command will plot the curve and draw the secant line from (0, 0) to (10, 1000). The **PlotStyle** section controls the way the two curves are plotted. The first part, **GrayLevel[0]** applies to the first curve in the list, $s(t)$. It will be drawn in black. The second part, **Thickness[0.015], GrayLevel[0.5]** applies to the second curve, the secant line. The line will be thickened and gray instead of black. Copy and edit the **Plot** command above, add secLine[t], put braces around the two functions, and add the **PlotStyle** section.

```
In[8]:=
    Plot[
        {s[t], secLine[t]}, {t,0,10},
        AxesLabel->{" t", "s(t)"},
        PlotLabel->"s(t) = t^3",
        PlotStyle->
            {{GrayLevel[0]},
            {Thickness[0.015], GrayLevel[0.5]}
            },
        Epilog->
            {PointSize[.03],
            Point[{0,0}],
            Point[{10,1000}]
            }
    ];
```

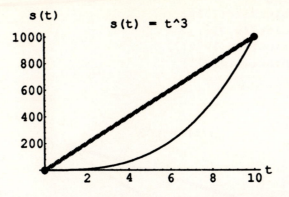

The slope of the secant line is just the average velocity the object has during the first 10 seconds of its motion. But the instantaneous velocity is also a slope. Since $v(t) = s'(t)$ and $s'(t)$ is the slope of the tangent line to this curve, the slope at any time t is equal to the instantaneous velocity at t. The problem we are considering is equivalent to that of finding a point on the curve where the tangent is parallel to the secant line.

In order to investigate this further, define a function that will give the tangent line to $y = s(t)$ at the point $t = c$. When a value of c is specified, this function gives the formula for the tangent line.

```
In[9]:=
        tanLine[t_,c_]:= s'[c](t - c) + s[c]
```

Let's add the tangent line at the point (8, 256) to the plot above. Copy and edit the preceding **Plot** command, adding **TanLine[t, c]** to the function list, and **Point[c, s[c]]** to the point list is the **Epilog** section. Set $c = 8$ before plotting.

```
In[10]:=
        c = 8

Out[10]=
        8


In[11]:=
        Plot[
          {s[t],secLine[t],tanLine[t,c]},
          {t,0,10},
          AxesLabel -> {" t", "s(t)"},
          PlotLabel -> "s(t) = t^3",
          PlotStyle->{
            {GrayLevel[0]},
            {Thickness[0.015],GrayLevel[0.5]}
            },
```

```
Epilog->{
   PointSize[.03],
   Point[{0,0}],
   Point[{10,1000}],
   Point[{c,s[c]}]
   }
];
```

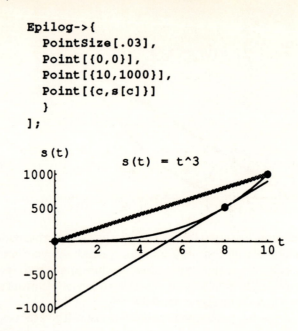

Practice Exercise 1: Draw a diagram like the one above for the average velocity over the interval from t = 2 to t = 8 and the instantaneous velocity at t = 4. Try giving **GrayLevel** and **Thickness** different values and see what happens.

Is there a point on the curve where the tangent line is parallel to the secant line? From the plot it looks like there should be. Let's find it. We know the slope of the secant line is $m = (s(10) - s(0))/(10 - 0) = 100$ ft/sec. The slope of the tangent line is $s'(t)$. So set $s'(t) = 100$ and solve for t.

```
In[12]:=
        Solve[s'[t]==100, t]
```

Out[12]=
$$\{\{t \to \frac{10}{\text{Sqrt}[3]}\}, \{t \to \frac{-10}{\text{Sqrt}[3]}\}\}$$

We need a decimal value, not a symbolic one, and we want the *t* value between 0 and 10.

```
In[13]:=
        c = N[10/Sqrt[3]]
```

Out[13]=
        5.7735

Note that this is the same value you found by setting the instantaneous velocity equal to the average velocity.

Now copy the **Plot** command and enter it. Since the value of **c** is now 5.7735, the graph will be plotted with the tangent line at $t = 5.7735$.

```
In[14]:=
        Plot[
          {s[t],secLine[t],tanLine[t,c]},
          {t,0,10},
          AxesLabel -> {" t", "s(t)"},
          PlotLabel -> "s(t) = t^3",
          PlotStyle->{
            {GrayLevel[0]},
            {Thickness[0.015],GrayLevel[0.5]}
            },
          Epilog->{
            PointSize[.03],
            Point[{0,0}],
            Point[{10,1000}],
            Point[{c,s[c]}]
            }
          ];
```

The tangent line at $t = 5.7735$ is parallel to the secant line. That is, at time $t = 5.7735$, the instantaneous velocity of the object is equal to its average velocity over the interval $0 \le t \le 10$.

Practice Exercise 2: At what time between $t = 2$ and $t = 8$ will the instantaneous velocity equal the overall average velocity for the time interval? Draw a plot for this time interval like the one above. (*Ans. $t = 5.2915$ sec*)

The observations which have been made here for distance and its derivative, velocity, generalize to a very large class of functions. The general principle is stated as follows.

The Mean Value Principle

> The average rate of change of a function over an interval must be equal to its instantaneous rate of change at some point in the interval.

Here is an example of a function, $f(x) = x^3 - 6x^2 - x + 30$, $-3 \le x \le 6$, to which the principle applies:

```
In[15]:=
        Clear[m, secLine, t, x];
        f[x_]:= x^3 - 6x^2 - x + 30
```

```
In[16]:=
        Plot[
          f[x], {x,-3,6},
          PlotLabel->"f[x] = x^3 - 6x^2 - x +30",
          AxesLabel->{"x", ""}
          ];
```

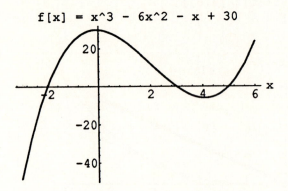

The average rate of increase in this function over the interval from -3 to 6 can be found as follows:

```
In[17]:=
        m = (f[6] - f[-3])/(6 - (-3))
```

```
Out[17]=
        8
```

In other words, the function increases on the average at a rate of 8 units for each unit increase in x. This average is the slope of the line joining the endpoints of the graph.

```
In[18]:=
        secLine[x_]:= m (x + 3) + f[-3]
```

Copy, edit, and enter the previous **Plot** command to produce the next one.

```
In[19]:=
        Plot[
          {f[x], secLine[x]}, {x,-3,6},
          AxesLabel->{" x", ""},
          PlotLabel->"f(x) = x^3 - 6x^2 - x +30",
          PlotStyle->{
            {GrayLevel[0]},
            {Thickness[0.015], GrayLevel[0.5]}
            },
          Epilog->{
            PointSize[.03],
            Point[{-3,f[-3]}],
            Point[{6,f[6]}]
            }
          ];
```

$$f(x) = x\text{\textasciicircum}3 - 6x\text{\textasciicircum}2 - x + 30$$

At what points is the instantaneous slope equal to the average slope? Solve for x.

```
In[20]:=
        N[Solve[f'[x]==m, x]]
```

```
Out[20]=
        {{x -> 4.64575}, {x -> -0.645751}}
```

This time there are two solutions within the interval. Plot to verify that the tangent line really is parallel at those points.

```
In[21]:=
        c1 = 4.64575
        c2 = -0.645751
```

```
Out[21]=
        -0.645751
```

```
In[22]:=
        Clear[tanLine]
        tanLine[x_,c_]:= f[c] + f'[c](x - c)
```

Copy and edit the previous **Plot** command to produce this one.

```
In[23]:=
        Plot[
          {f[x], secLine[x], tanLine[x,c1],
            tanLine[x,c2]},
          {x,-3,6},
          AxesLabel->{" x", ""},
          PlotLabel->"f(x) = x^3 - 6x^2 - x +30",
          PlotStyle->{
            {GrayLevel[0]},
            {Thickness[0.015],RGBColor[0,1,0]},
            {GrayLevel[0]},
            {GrayLevel[0]}
            },
          Epilog->{
            PointSize[.03],
            Point[{-3,f[-3]}],
            Point[{6,f[6]}]
            }
          ];
```

f(x) = x^3 - 6x^2 - x + 30

The Mean Value Principle Is Not True for All Functions Over All Intervals

As compelling and useful as the principle is, it does not, as you shall soon see, apply to all functions. In order to use the principle intelligently, you need to have a clear notion of the circumstances in which it applies and those in which it doesn't.

Consider the rational function $y = 1/x$ over the interval from $-.5$ to 1.

```
In[24]:=
        Clear[f,x];
        f[x_]:= 1/x

In[25]:=
        m = (f[1.0] - f[-0.5])/(1.0 + 0.5)

Out[25]=
        2.

In[26]:=
        Clear[secLine];
        secLine[x_] = f[-0.5] + m(x + 0.5)

Out[26]=
        -2. + 2. (0.5 + x)

In[27]:=
        Plot[
          {f[x], secLine[x]}, {x,-0.5,1.0},
          PlotRange->{-20,20},
          PlotStyle->{
            {GrayLevel[0]},
            {Thickness[0.015],RGBColor[0,1,0]},
            },
          Epilog->{
            PointSize[.03],
            Point[{-0.5,f[-0.5]}],
            Point[{1,f[1]}]
            }
          ];

Power::infy:
                              1
        Infinite expression - encountered.
                              0.
```

```
Power::infy:
                        1
    Infinite expression -- encountered.
                        0.
```

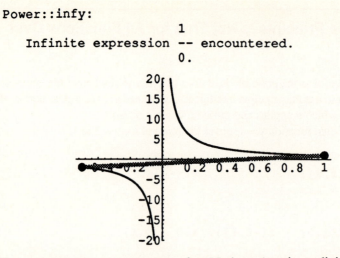

You may get some error messages when *Mathematica* tries to divide by 0 at $x = 0$.

It does not appear that the tangent line at any point will be parallel to the secant line drawn on this curve. In fact, the derivative is always negative, but the average slope is $m = 2$. If you try to solve for the points as we did for the previous function, you do not get a real solution.

```
In[28]:=
        Solve[f'[x]==2,x]
```

```
Out[28]=
                    I                    -I
        {{x ->  ---------- },  {x ->  ---------- }}
                  Sqrt[2]               Sqrt[2]
```

The Mean Value Principle failed here because the function, $1/x$, has a discontinuity in the interval over which we were averaging it. Had the graph joined the endpoints in a nice smooth curve, no such problem would have arisen. As a matter of fact, when we consider the same function over an interval that does not contain zero, the Mean Value Principle does hold.

Practice Exercise 3: Show that the Mean Value Principle does hold for $f(x) = 1/x$ on the interval $0.5 \le x \le 2.0$.

The curve $y = h(x) = x^{(2/3)}$ illustrates another possible difficulty in applying the Mean Value Principle. Consider this function on the interval $-1 \le x \le 2$. There is a technical difficulty in defining this function in *Mathematica*. Since it has a cube root, there are actually three values for $x^{(2/3)}$ at any particular x. Two of these values are complex and one is real. We are only interested in the real values, but when x is negative, *Mathematica* gives one of the complex values. But by defining the function in the form $(x^2)^{(1/3)}$, we can get *Mathematica* to return a real value for all x.

```
In[29]:=
        h[x_]:=(x^2)^(1/3)
```

```
In[30]:=
        m = N[(h[2] - h[-1])/(2 + 1)]

Out[30]=
        0.1958

In[31]:=
        secLine[x_]:= h[-1] + m(x + 1)

In[32]:=
        Plot[
          {h[x], secLine[x]}, {x,-1,2},
          PlotStyle->{
            {GrayLevel[0]},
            {Thickness[0.015],RGBColor[0,1,0]},
            },
          Epilog->{
            PointSize[.03],
            Point[{-1,h[-1]}],
            Point[{2,h[2]}]
            }
          ];
```

The slope of the secant line between $x = -1$ and $x = 2$ is $m = 0.1958$, but there is no point between -1 and 2 where the tangent is parallel to the secant.

```
In[33]:=
        Solve[h'[x]==0.1958,x]

Out[33]=
        {{x -> 39.4719}}
```

Thus even though this function is continuous in the interval $-1 \le x \le 2$, the Mean Value Principle does not apply. This time the problem is that there is a point (only one!) where the derivative does not exist. If we plot the derivative, you see that it has a discontinuity at $x = 0$. (You may get some error messages as *Mathematica* tries to plot at $x = 0$.)

```
In[34]:=
        Plot[h'[x],{x,-1,2}];
```

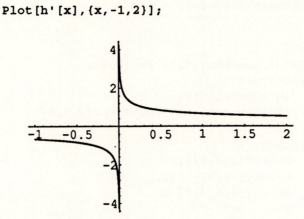

You can see from the examples you have studied that in applying the Mean Value Principle, you must be careful to avoid intervals over which the function is discontinuous or in which its derivative is undefined. These are, as it turns out, the only possible pitfalls in applying the principle. This fact is set out in an important mathematical result that is known as the Mean Value Theorem. It is quoted below as it appears in a standard calculus text.

The Mean Value Theorem

> If $y = g(x)$ is continuous at every point of the closed interval $[a, b]$ and differentiable at every point of its interior (a, b), then there is at least one number c between a and b at which
>
> $$(g(b) - g(a))/(b - a) = g'(c).$$

Code for the Practice Exercises

```
1. Clear[s,m,secLine,tanLine];
   s[t_]:= t^3;
   m = (s[8] - s[2])/(8 - 2);
   secLine[t_]:=s[2] + m(t - 2);
   tanLine[t_,c_]:= s[c] + s'[c](t - c);
   c = 4;
   Plot[
     {s[t], secLine[t], tanLine[t,c]},
     {t,2,8},
     AxesLabel -> {" t", "s(t)"},
     PlotLabel -> "s(t) = t^3",
     PlotStyle->{
       {GrayLevel[0]},
       {Thickness[0.015], GrayLevel[0.5]}
       },
     Epilog->{
       PointSize[.03],
       Point[ {2, s[2]} ],
       Point[ {8, s[8]} ],
       Point[ {c, s[c]} ] }
       }
     ];
2. Solve[s'[t]==m,t];
   N[%]

   c=5.2915;
   Plot[
     { s[t], secLine[t], tanLine[t,c] },
     {t,2,8},
     AxesLabel -> {" t", "s(t)"},
     PlotLabel -> "s(t) = t^3",
     PlotStyle->{
       {GrayLevel[0]},
       {Thickness[0.015], GrayLevel[0.5]}          },
     Epilog->{
       PointSize[.03],
       Point[ {2, s[2]} ],
       Point[ {8, s[8]} ],
       Point[ {c, s[c]} ]
       }
     ];
```

3.
```
Clear[f, secLine, tanLine];
f[x_]:=1/x;
m = (f[2.0] - f[0.5])/(2.0 - 0.5)
secLine[x_]:=f[0.5] + m(x - 0.5);
tanLine[x_,c_]:=f[c] + f'[c](x - c);

Solve[f'[x]==m,x]

c = 1;
Plot[
   {f[x], secLine[x], tanLine[x,c]},
   {x,0.5,2.0},
   PlotStyle->{
      {GrayLevel[0]},
      {Thickness[0.015],GrayLevel[0.5]}
      },
   Epilog->{
      PointSize[.03],
      Point[ {0.5, f[0.5]} ],
      Point[ {2.0, f[2.0]} ],
      Point[ {c,   f[c]}   ]
      }
   ];
```

Problems

1. For each of the following functions make and print a plot which verifies that the Mean Value Theorem is true for the function over the indicated interval.

   (a) $f(x) = x^3 - 2x^2$, $[-1,1]$
   (b) $f(x) = (x - 1)^{(1/2)}$, $[1,5]$
   (c) $f(x) = (x^3 + 8)/(x - 2)$, $[0,1]$

2. For the function in 1c above, find an interval over which the Mean Value Theorem does not apply. Print out a graph of the function over this interval along with a written analysis of the difficulty.

3. Explain why the following statement is true for any function, $F(x)$, to which the Mean Value Theorem applies: If the graph of $y = F(x)$ crosses the x-axis at x_1 and x_2, then $F'(x)$ is zero for an x-value between x_1 and x_2.

4. Find a function which crosses the x-axis at $x = 0$ and at $x = 4$, but whose derivative is never 0 in the interval $0 \le x \le 4$. Print its graph and give a written analysis of the function's behavior.

10

Assignments and Definitions in *Mathematica*

From time to time in the past nine explorations we have promised an explanation of a peculiarity of *Mathematica* syntax. In this exploration we are going to make good on those promises.

Equality

Why does *Mathematica* have so many different "equal" signs? We have encountered three of them: == (double equals), := (colon equal), and = (equal). Ordinary paper and pencil mathematics gets along quite well with only one equal sign. Why does *Mathematica* need three? (Actually there are four: === is another type of equality, but it will not be used in any of our work, so we need not concern ourselves with it here.) The fact of the matter is that in ordinary mathematical work we use the equal sign to mean different things. If we write "$0.5 = 1/2$" we mean that the number on the right is the same as the number on the left. On the other hand, "$x^2 = 4$", means something quite different. It is a way of saying, the numbers x for which the statement "$x^2 = 4$" is true. When we write an algebraic rule like $(a + b)^2 = a^2 + 2ab + b^2$, we mean that the expression on the left gives the same result as the expression on the right for all values of a and b. We are human beings; we learn to interpret the equal sign from the context of the statement. This is a much more difficult feat for a computer, so the designers of *Mathematica* used a different sign for each relationship between variables and numbers.

THE LOGICAL EQUALS, ==

The symbol == makes a statement about the expressions on either side of it; the statement is usually true or false. The statement is judged true if when the expression on the left-hand side is evaluated, it gives the same *value* as the expression on the right-hand side. Otherwise the statement is false. Following is a little example.

```
In[1]:=
        Clear[x,y,z];
        x = 2;
        y = 16;
        x == 1
```

```
Out[1]=
        False
```

The first three commands assign the value 2 to **x** and the value 16 to **y** but leave **z** undefined. *Mathematica* responds to the statement **x == 1** with False, because the value assigned to **x** is not 1. Try **x == 2**.

```
In[2]:=
        x == 2
```

```
Out[2]=
        True
```

Try a few more statements.

```
In[3]:=
        x^3 == 8
```

```
Out[3]=
        True
```

```
In[4]:=
        x == y
```

```
Out[4]=
        False
```

```
In[5]:=
        x^3 == y/2
```

```
Out[5]=
        True
```

```
In[6]:=
        x == z
```

```
Out[6]=
        2 == z
```

With one exception, the result is either `True` or `False`. *Mathematica* has evaluated both sides of the statement and returned `True` when the values were equal or `False` when they were not equal. The exception is `x == z`. In this case `z` does not have a value, so the statement is neither true nor false. Essentially, *Mathematica* is saying that `x == z` could be true, or it could be false, depending in what value is assigned to `z`.

One very important use of the double equals symbol is in the `Solve` command. When you give a command like:

```
In[7]:=
        Clear[x]
```

```
In[8]:=
        Solve[6 - 7*x + x^3 == 0, x]
```

```
Out[8]=
        {{x -> -3}, {x -> 2}, {x -> 1}}
```

you are looking for the values of **x** that make `6 - 7*x + x^3 == 0` a true statement. Note that the expression itself is neither true nor false, because it contains a variable that does not have a value.

```
In[9]:=
        6 - 7*x + x^3 == 0
```

$$
Out[9]= \\
6 - 7\,x + x^3\ == 0
$$

But, when we evaluate the expression at a particular value of **x**, then it becomes either a true statement or a false one.

```
In[10]:=
        6 - 7*x + x^3 == 0/.x->1
```

```
Out[10]=
        True
```

```
In[11]:=
        6 - 7*x + x^3 == 0/.x->0
```

```
Out[11]=
        False
```

The output from `Solve` is exactly those assignments to **x** which make the statement true.

```
In[12]:=
        6 - 7*x + x^3 == 0/.
            {{x -> -3}, {x -> 2}, {x -> 1}}

Out[12]=
        {True, True, True}
```

ASSIGNMENT STATEMENTS

The other two "equal" signs are not equality statements at all; they are instructions that tell *Mathematica* to *assign* a value to a variable. That means that the left-hand side must be the name of a variable; it cannot be an expression. The right-hand side may be any valid *Mathematica* expression.

There is an important difference in the way the two assignment commands behave. The plain "equal" sign (=) evaluates the right-hand side immediately, and assigns that value to the variable on the left. Whenever you refer to that variable you always get that same value. In contrast, the colon equal sign (:=) does *not* evaluate the right-hand side; instead it saves the expression in symbolic form. The expression is not evaluated until you refer to the variable. And each time you refer to the variable, the expression is reevaluated. If any of the variables in the expression were changed in between references, you get a different value for the expression. It may seem like a small difference, but it can drastically change the result of a calculation. The following examples will help make the difference clear.

Here are the initial assignments for **x**, **y**, and **z**:

```
In[13]:=
        Clear[x,y,z]

In[14]:=
        x = 1

Out[14]=
        1

In[15]:=
        y = x^2

Out[15]=
        1

In[16]:=
        z:= x^2
```

After these three assignment statements, the variable **y** has the value 1, because the expression x^2 was evaluated with **x** = **1** and assigned to **y**. But **z** has the value x^2, not 1. When you use **z** in an expression, **x** is set to its current value, making it seem as if that **y** and **z** have the same value.

```
In[17]:=
        Y
```

```
Out[17]=
        1
```

```
In[18]:=
        z
```

```
Out[18]=
        1
```

They even pass the equality test.

```
In[19]:=
        y == z
```

```
Out[19]=
        True
```

But look what happens when you change the value of **x**.

```
In[20]:=
        x = 2
```

```
Out[20]=
        2
```

```
In[21]:=
        Y
```

```
Out[21]=
        1
```

```
In[22]:=
        z
```

```
Out[22]=
        4
```

The variable **y** still has the value 1 that it was assigned. The variable **z** is now evaluated with **x = 2** and so evaluates to 4. Now **y** and **z** fail the equality test.

```
In[23]:=
        y == z
```

```
Out[23]=
        False
```

Practice Exercise 1: Try clearing **x** and see what happens to **y** and **z**. (*Ans.* **y = 1**; **z = x^2**)

Keeping Track of Variable Assignments

It is very easy to lose track of how a variable is defined. Fortunately, you can ask *Mathematica* to remind you by using a question mark, followed by the variable name.

```
In[24]:=
        ?y
```

```
Global'y

y = 1
```

The variable y is defined to be 1.

```
In[25]:=
        ?z
```

```
Global'z

z := x^2
```

The variable z is defined to be x^2.

When Do You Use Which Assignment Symbol?

When you are doing simple calculations it may not matter much whether you use = (equal) or := (colon equal). However, for more complex operations, you need to be careful. Usually, to assign a value to a variable, you should use = (equal). To define a function, use := (colon equal). When you do something more complicated, like setting a variable t equal to an expression which itself contains other variables, think about the variables in the expression on the right side. The values of these variables may change. If you want the new variable (the one on the left) to reflect those changes, then use :=. Use the plain = if you do not. In other words, use := if you are giving a command that will later be used to calculate a value, and use = if you are using the command to determine the *final* value of your expression now.

Here is one situation in which you need to make a definition using = instead of :=. Suppose you have a function, say $f(x) = x^3$, and want to define the function $g(x)$ to be its derivative. If you write

```
In[26]:=
        Clear[f,x];
        f[x_]:= x^3
```

```
In[27]:=
        g[x_]:= D[f[x],x]
```

```
In[28]:=
        g[x]
```

```
Out[28]=
        3 x²
```

it looks fine. But try to evaluate *g* at 1.

```
In[29]:=
        g[1]
```

```
General::ivar: 1 is not a valid variable.
```

```
Out[29]=
        D[1, 1]
```

When *Mathematica* tried to evaluate `g[1]`, it first substituted 1 for **x** in the expression `D[f[x], x]`, getting `D[f[1], 1]`, which is nonsense.

Now, try defining the derivative with = instead of :=.

```
In[30]:=
        h[x_] = D[f[x],x]
```

```
Out[30]=
        3 x²
```

```
In[31]:=
        h[1]
```

```
Out[31]=
        3
```

This time it works fine.

Note: Also see Problem 1 at the end of this Exploration.

Practice Exercise 2: Look at the definitions stored by *Mathematica* for **g** and **h**. Explain why the one for **h** works correctly.

More on Assignments

Whenever you refer to a variable, a place for that variable is made in *Mathematica*'s definition table. Sometimes new definitions replace old ones; but in other circumstances the new definition is just added on to the table in a way that leaves the other definitions unchanged. Sometimes this is appropriate, but occasionally when the user has not kept track of the variables used in the definitions, strange bugs are created that can be very hard to find. Suppose for example that you want to define the function $f(x) = x^3$, but you forget the underscore after the **x**. And suppose further that you are in the middle of a long session and you are very tired, so you also forget that an hour ago you assigned **x** the value 5. What you have made for yourself is a mess.

```
In[32]:=
        x = 5
        Clear[f]
```

Type the function definition incorrectly:

```
In[33]:=
        f[x]:= x^2
```

Then correct it:

```
In[34]:=
        f[x_]:= x^2
```

This function seems to work perfectly. Try finding $f(4)$, $f(5)$, and $f(6)$ to convince yourself. But later suppose you try to change **f** to a different function.

```
In[35]:=
        f[x_]:= 1/x
```

Try some values.

```
In[36]:=
        f[2]
```

```
Out[36]=
        1
        -
        2
```

```
In[37]:=
        f[4]
```

```
Out[37]=
        1
        -
        4
```

It works perfectly, at least until you try $f(5)$

```
In[38]:=
        f[5]
```

```
Out[38]=
        25
```

How could **f [5]** be 25? Even if you type in and enter the correct function $f(x) = 1/x$ again, you still get this error. The explanation lies in the way *Mathematica* records its definitions. Look at what it lists for **f**.

```
In[39]:=
          ?f
```

```
Global'f
```

```
f[5]:= x^2
```

```
f[x_]:= 1/x
```

Mathematica thinks that **f** has two parts. If the argument is 5, then *Mathematica* returns x^2 or 25. That's because you entered **f [x] := x^2** when **x** had the value 5. *Mathematica* has not forgotten this instruction. For all the other values of x it understands that you mean $1/x$. Look at **f [5]** after **x** is changed to 2.

```
In[40]:=
          x=2
```

```
Out[40]=
          2
```

```
In[41]:=
          f[5]
```

```
Out[41]=
          4
```

Why did **f [5]** return 4?

How do you get out of this dilemma? If you **Clear** the function symbol **f** of any old definitions before you define new ones, you should be OK.

```
In[42]:=
          Clear[f]
```

```
In[43]:=
          ?f
```

```
Global'f
```

Now, the variable **f** has no definitions; so when we give a new one, there will be nothing to get in the way.

Failing to **Clear** variables can cause other problems as well. Suppose again that you made the assignment **x = 5** previously but have forgotten about it.

```
In[44]:=
          x = 5
```

```
Out[44]=
        5
```

```
In[45]:=
        Solve[6 - 7*x + x^3 == 0,x]
```

```
Out[45]=
        {}
```

The equation seems to have no solutions! But this is the same equation we solved earlier and found three roots. What is going on? Well, since **x = 5**, the expression **6 - 7*x + x^3** evaluates to 96.

```
In[46]:=
        6 - 7*x + x^3
```

```
Out[46]=
        96
```

And since 96 is not equal to 0, there are no numbers for which the statement **6 - 7*x + x^3 == 0** is true. Again the problem is solved by clearing a variable, this time **x**.

Incidentally, you may have noticed that **x** does not need to be cleared before you make a function definition. That is because *Mathematica* knows that when you use an **x_** in the definition form **f[x_]:=...**, you want any **x**'s on the right-hand side to remain symbolic. It does not evaluate them.

Information about Commands

The question mark command that we used to see definitions can also be used to find out about *Mathematica*'s built-in functions. To find the syntax of a command, for example, type a question mark followed by the name.

```
In[47]:=
        ?Plot
```

```
Plot[f, {x, xmin, xmax}] generates a plot of f
        as a function of x from xmin to xmax.
        Plot[{f1, f2, ...}, {x, xmin, xmax}] plots
        several functions fi.
```

The output describes the **Plot** command.

You can even get a list of options by using two question marks.

```
In[48]:=
        ??Plot

Plot[f, {x, xmin, xmax}] generates a plot of f
        as a function of x from xmin to xmax.
        Plot[{f1, f2, ...}, {x, xmin, xmax}] plots
        several functions fi.

        Attributes[Plot] = {HoldAll, Protected}

        Options[Plot] =
          {AspectRatio -> GoldenRatio^(-1),
          Axes -> Automatic, AxesLabel -> None,
          AxesOrigin -> Automatic,
          AxesStyle -> Automatic,
          Background -> Automatic,
          ColorOutput -> Automatic,
          Compiled -> True,
          DefaultColor -> Automatic, Epilog -> {},
          Frame -> False, FrameLabel -> None,
          FrameStyle -> Automatic,
          FrameTicks -> Automatic,
          GridLines -> None, MaxBend -> 10.,
          PlotDivision -> 20., PlotLabel -> None,
          PlotPoints -> 25, PlotRange -> Automatic,
          PlotRegion -> Automatic,
          PlotStyle -> Automatic, Prolog -> {},
          RotateLabel -> True, Ticks -> Automatic,
          DefaultFont :> $DefaultFont,
          DisplayFunction :> $DisplayFunction}
```

Although this does not explain what the options do, it will remind you of their proper form and spelling.

A Problem with Using Functions Defined in Packages

Packages usually contain functions that extend the basic operations of *Mathematica*. You must load the package before you can use a function from it. It is easy to forget this preliminary step and to type and enter the function's name before loading the package. This creates a problem that even **Clear** will not solve. What happens is this. When you refer to a name that has not been defined (in this case, the name of the function), *Mathematica* creates a place for that name in the definition table. The name is there, but there is no definition for it. When you read in the package, the names of the functions from that package are put into a special place and do not replace those that are already in the definition table. When you try to use the function, *Mathematica* finds the original one, the one with no definition, instead of the one you really want from the package. Using **Clear** will not help, since the function does not have a definition anyway. What you need to do is to take the *name* out of the definition table completely, which is done with the **Remove** command.

Here is an example of a function name being placed in the definition table and taken out.

```
In[49]:=
        Remove[fun] (* Make sure fun is not in the defi-
        nition table *)
```

```
In[50]:=
        Clear[x]
```

```
In[51]:=
        ?fun
```

```
Information::notfound: Symbol fun not found.
```

There is no symbol **fun** in the definition table. Suppose you use the name by mistake.

```
In[52]:=
        z = fun[x]
```

```
Out[52]=
        fun[x]
```

Since **fun** is not defined, the assignment statement gives output that just repeats the input. If you expected the function to be defined, say because you loaded a package that was supposed to contain its definition, then this is your first clue that something is wrong. (If you had used := (colon equal) instead of = you would not even have gotten that clue, since an assignment using := produces no output.). So, ask *Mathematica* what it knows about **fun**.

```
In[53]:=
        ?fun
```

```
Global'fun
```

All it knows is that **fun** is in the definition table. Now clear **fun**.

```
In[54]:=
        Clear[fun]
        ?fun
```

```
Global'fun
```

Clear did not take **fun** out of the table. So, try **Remove**. Make sure that the **Remove** command and **?fun** are in different cells.

```
In[55]:=
        Remove[fun]

In[56]:=
        ?fun
```

Information::notfound: Symbol fun not found.

Remove did it; **fun** is no longer in the table of definitions.

Problems

1. We saw above that setting **g[x_] := D[f[x], x]** does not allow g[1] to be evaluated correctly. In *Mathematica* the derivative can also be invoked using prime notation, **f'[x]**. Does **g** evaluate correctly if you make the assignment **g[x_] := f'[x]**? Make some experiments and write an explanation of your conclusion.

2. What is the result of the following commands?

   ```
   x = 5
   f[x] := x^2
   f[x_] := x^2
   x = 3
   ```

   After entering the commands, find $f(3)$, $f(4)$, $f(5)$, and $f(6)$. Write a description of what happens and explain why it happens.

3. Give a sequence of commands so that $f(x)$ is defined in such a way that when you evaluate $f(2)$ the output is 8, but when you evaluate $f(x)$ for any other x value you get is $x + 1$.

11

Sines and Cosines

Preliminaries

In this section you will be using sequences of graphs and animations to help you to understand the behavior of sine and cosine curves. You will create the frames for many of these animations yourself. In a few cases the frames have already been drawn for you and are stored as a *Mathematica* package on your *Exploring Calculus* disk. We'll get to these animations in a bit, but first take a look at the basic sine curve.

```
In[1]:=
    Plot[
       Sin[x], {x, -4Pi, 4Pi},
         Ticks->{{-4Pi,-2Pi,0,2Pi,4Pi},
                {-1,1}},
         AxesLabel->{"x", "sin(x)"}
       ];
```

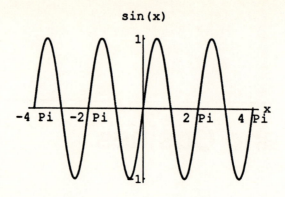

The plot displays the important features of the sine function: the values of sin(x) oscillate back and forth between −1 and +1 and a complete cycle of the curve is displayed as *x* increases in value from 0 to 2π. Recall the technical language: The curve has an *amplitude* of 1 and a *period* of 2π .

The Sine Curve

The characteristics of the sine curve come from the definition of the sine function. Look at the diagram below. The sine of the angle *x* is defined to be the second coordinate of the point P. Note that we will be using radian measure for the angle *x*. As the angle increases from zero to 2π, the value of P's second coordinate changes. Hence sin(x) increases from zero to one, decreases to negative one, and then increases again to zero. As the angle increases further, the same values for sin(x) are repeated over and over again, causing the periodic nature of the sine function. When the values for *x* are negative, the angles are generated in a clockwise direction. Their sines display the same cyclical character described above for positive values of *x*.

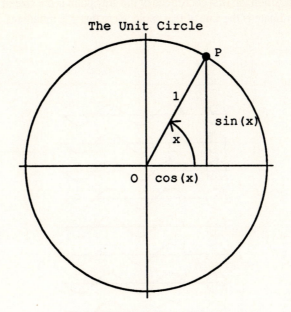

An Animation of the Sine Function

In this section we will look at an instructive moving picture which shows dynamically how the sine curve is shaped by the unit circle definition of the sine function. Below are the frames of an animation of the sine. As the black point moves around the circle, its directed distance from the horizontal axis gives the sine of the central angle formed at the origin by the horizontal axis and the line joining the black point to the origin. The gray point on the curve to the right moves through the corresponding positions on the sine curve. In each position the first coordinate of the gray point is the size of the angle x and the second coordinate is the sine of that angle. Note that x is in radians, not degrees.

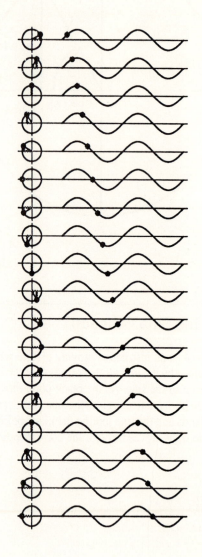

These frames can be drawn in your notebook by a package called **SineAn.m**. Just load the package by typing and entering

```
<<SineAn.m
```

Mathematica may respond with a dialog box asking you to find the file; if so, you need to know where to locate it. It is on the *Exploring Calculus* disk in the **Packages** file folder under **TrigAn**. If you are unfamiliar with the procedure for finding a package, see Exploration Five.

Once the frames are on your screen, you can easily animate the display. All you have to do is select the cell containing the frames, pull down the **Graph** menu and release on **Animate Selected Graphics**. The speed and direction of the animation can be controlled using the horizontal scroll bar. For detailed instructions on how this is done see Exploration Five.

Animations Take Memory, Lots of It

Animations are fun, and they can be instructive, but they require a great deal of memory because of the large number of graphs required and so can cause *Mathematica* to crash. This means you must exercise some care in running them. Here are some suggestions:

1. Save your file before executing any command that will produce a number of plots. This way if the system bombs, you haven't lost your work.

2. When you are done with a sequence of plots, Clear them from your notebook in order to free up the memory they use.

Practice Exercise 1: Clear the frames for the sine animation and run the cosine animation. The frames will be drawn in your notebook by the package <<**CosineAn.m**. What is the period and amplitude of the cosine curve? (*Ans.* 2π, 1)

Changing the Amplitude of the Sine Curve

The cyclical nature of the basic sine curve is a characteristic that is shared by all sine curves. For example, look at this variation on the sine curve: $y = 3\sin(x)$ plotted over the interval $-6\pi \le x \le 6\pi$. It is drawn below with $\sin(x)$.

```
In[2]:=
    Plot[
      {3Sin[x], Sin[x]}, {x,-4Pi, 4Pi},
      PlotStyle->
        {Thickness[.01], Thickness[.004] }
      ];
```

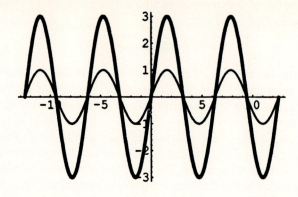

The graph of $y = 3\sin(x)$ is the heavier line. You can see that it has the same period as the sine curve but has three times the amplitude.

Creating Animations Using the Do Command

It is easy in *Mathematica* to create your own animations, and the result can be both entertaining and instructive. All you need is a command which will draw one plot after another. Once you have the pictures, the work is over. The **Do** command does the trick. Below you see an example of **Do** set up to generate a sequence of sine curves of the form $y = a\sin(x)$, $a = -4, -3, \ldots, 5$. The output is displayed here in compressed form. When you enter the command in your notebook, your plots will be drawn vertically one after another in perfect form for animation.

```
Do[
        Plot[
          a*Sin[x], {x, -4Pi, 4Pi},
          PlotRange->{-4.5,4.5},
          PlotLabel->"a = "<>ToString[a]
          ],
        {a,-4,5}
        ]
```

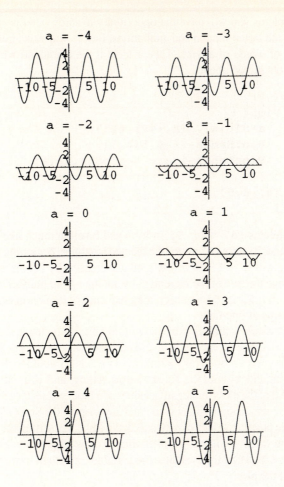

Run the animation for a vivid picture of the effect of *a* on the shape of the sine curve. Pause the animation by clicking on the pause button at the lower left corner of the window. Then press and hold on the horizontal scroll bar box. Move the bar left and right slowly and observe the changes in the plot. All of the plots have the same period, only the amplitude changes from one to the next. It begins at 4, decreases to 0, and then increases to 5 as the value of *a* changes.The amplitude of each of the sine curves is |*a*|.

In the problems you will have the opportunity to study other properties of the sine function using the **Do** command and animations. Before you strike out on your own, we have a couple of words of advice. Take a look at the command which generated the sequence of plots shown above.

```
Do[
  Plot[
    a*Sin[x], {x,-4Pi,4Pi},
    PlotRange->{-4.5,4.5},
    PlotLabel->"a = "<>ToString[a]
    ],
  {a,-4,5}
  ]
```

It contains an iterator **{a, -4, 5}**, in form and function much like the iterator in the **Table** command. It causes **Plot** to be executed ten times, beginning with the curve for **a = -4**, ending with **a = 5**, and with **a**'s value increasing by 1 from one plot to the next. You could change the size of the increment by adding a third number to the iterator. For example, **{a, -4, 5, 0.5}** would create a sequence of 19 plots in which the amplitude increased in jumps of one-half.

The **PlotRange** setting is extremely important. If you leave it out, *Mathematica* will use its default range for each plot, and so each sine curve will have a different vertical scale. In fact, the curves plotted for $y = \sin(x)$ and $y = 2\sin(x)$ would look exactly the same. Try the **Do** command without the **PlotRange** setting, and you will see what we are talking about. All of the plots for negative values of *a* look exactly the same. The only way you can tell they are different plots is to carefully inspect the vertical scales. The plots generated by the positive values of *a* also look the same. By setting the *y*-range in all the plots to the same value, we forced *Mathematica* to draw all the curves to the same scale.

The function **ToString** evaluates the variable *a* and then formats it, so that it can be used to label the plot.

Problems

1. In the function $y = a \sin(x)$ the letter *a* is called a *parameter*. For any particular sine curve, *a* would have a numerical value, while *x* and *y* remain variables. Sequences of plots and animations are very useful tools in studying variations in parameters, and you will be using them in this assignment to study the parameters *b*, *c*, and *d* of the general sine curve

$$y = a \sin(\, b(x - c)\,) + d.$$

(a) Use **Do** to generate a sequence of plots in which the value of the parameter d in $y = \sin(x) + d$ varies. Try $d = -2, -1, 0, 1$, and 2, for example, and use a **PlotRange** setting that runs from -3.5 to 3.5. Print your plots and write a short paragraph explaining how changes in d effect the plot. The parameter d is called the *displacement*.

(b) Make a similar analysis of the effect of the parameter b in $y = \sin(b*x)$. Try b ranging from 1 to 4 in 8 steps.

(c) Analyze the effect of the parameter c in $y = \sin(x - c)$. This parameter is called *phase shift*. Try c ranging from 0 to π in 6 (or more) steps.

2. Print out a plot of several periods of each of the following sine curves: a sine curve with

(a) amplitude $= 1/2$, period $= \pi/3$, phase shift $= \pi/4$, and an upward displacement of 3.

(b) amplitude $= 10$, period $= 6\pi$, phase shift $= -\pi$, and downward displacement of -2.

3. Plot each of the following functions and use the plot to determine amplitude and period.

(a) $y = 3\cos(2x)$

(b) $y = \cos^2(x)$

(c) $y = \sin(x) + \cos(x)$

4. For each of the curves in Problem 3, use the plots to estimate values for a, b, c, and d so that $y = a\sin(b(x - c)) + d$. Check your values by plotting the new equation and the original one together in the same graph. Repeat until you are satisfied that your values are correct. Print your last plot.

12

Derivatives of Sines and Cosines

The Derivative of the Sine

The graph of the sine function $y = \sin(x)$ is a smooth curve and so has a slope at every point.

```
In[1]:=
    Plot[
      Sin[x], {x,-2Pi,2Pi},
      AxesLabel->{"x", "y = sin(x)" },
      Ticks->{{-2Pi,-Pi,0,Pi,2Pi}, {-1,1}}
      ];
```

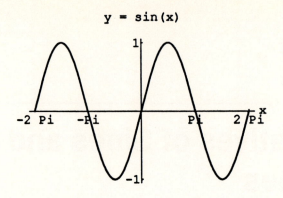

$$y = sin(x)$$

The derivative of sin(x) is the expression that predicts these slopes in terms of x. In this exploration we'll be trying to figure out what that expression should be. Since it is easier to think about the slope at a particular point, let's begin by estimating a few sine slopes.

The slope at a point (x, sin(x)) on the sine curve can be approximated by the difference quotient

$$\frac{Sin(x + dx) - Sin(x)}{dx}$$

when the value of dx is small. Here is an approximation to the tangent slope at the point (0, sin(0)) with $dx = 0.1$:

```
In[2]:=
        (Sin[0 + 0.1] - Sin[0])/0.1

Out[2]=
        0.998334
```

We can obtain a better approximation by taking a smaller value for dx.

```
In[3]:=
        (Sin[0 + 0.01] - Sin[0])/0.01

Out[3]=
        0.999983
```

Looks like the tangent slope at 0 is probably 1.

Let's try to find the slope at a different point, say when $x = 3\pi/4$. Calculate the difference quotient for $dx = 0.01$.

```
In[4]:=
        x = N[3Pi/4];
        (Sin[x + 0.01] - Sin[x])/0.01
```

Out [4] =

-0.710631

Try some smaller *dx* values.

In[5]:=

```
(Sin[x + 0.001] - Sin[x])/0.001
```

Out [5] =

-0.70746

In[6]:=

```
(Sin[x + 0.0001] - Sin[x])/0.0001
```

Out [6] =

-0.707142

Estimated to three decimal places, this slope is probably –.707. We can test the estimate by plotting the line through $(3\pi/4, \mathrm{Sin}(3\pi/4))$ with slope –0.707 on the sine curve. If the line doesn't look tangent to the curve, we have a problem. If it does, it is likely that our estimate is a good one.

In[7]:=

```
Clear[x,y];
y = -0.707(x-3Pi/4) + Sin[3Pi/4];
Plot[{Sin[x],y}, {x,0,Pi}];
```

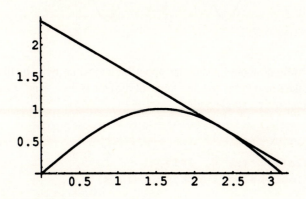

This looks pretty good.

The general difference quotient

$$\frac{\mathrm{Sin}(x + dx) - \mathrm{Sin}(x)}{dx}$$

will give a good estimate to the slope of the sine curve at $(x, \sin(x))$ as long as we use a small enough value for *dx*. This notion inspires a clever idea for investigating the sine's derivative. Write the difference quotient with a small number, say 0.01, substituted for *dx*.

$$\frac{\text{Sin}(x + 0.01) - \text{Sin}(x)}{0.01}$$

Note that it is a function of x. For any particular value of x it will produce a number close to the sine's slope at that value of x. This means that the difference quotient's values will be very close to those produced by the sine's derivative. So, if we graph the difference quotient function we will get a plot very close to the graph of the sine's derivative. Let's do it and see what we what the curve looks like.

```
In[8]:=
        Plot[
          (Sin[x + 0.01] - Sin[x])/0.01,
          {x,-2Pi,2Pi},
          Ticks->{{-2Pi,-Pi,0,Pi,2Pi}, {-1,1}}
          ];
```

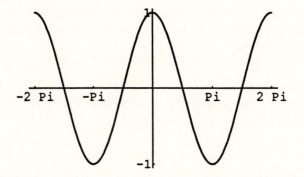

That curve has amplitude about 1 and period about 2π. It looks an awful lot like the cosine curve. Does it make sense to say the derivative of the sine is the cosine? Compare plots of the two functions over a single period.

```
In[9]:=
        Plot[
          Sin[x], {x, 0, 2Pi},
          PlotLabel->"function, sin(x)"
          ];
        Plot[
          Cos[x], {x, 0, 2Pi},
          PlotLabel->"derivative?, cos(x)"
          ];
```

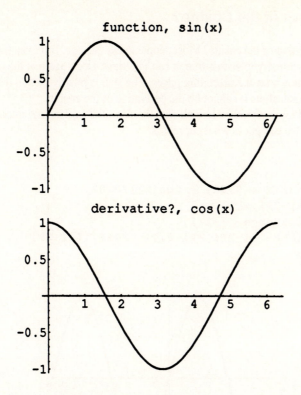

function, sin(x)

derivative?, cos(x)

Notice that the cosine is zero for those values of x where the sine curve attains a maximum or minimum and further, that the cosine curve is positive for exactly those values of x where sine rises and is negative where the sine falls. The cosine function does seem to correctly predict the direction of the sine curve.

Test this guess with a limit command.

```
In[10]:=
        Limit[(Sin[x + dx] - Sin[x])/dx, dx -> 0]

General::load: Loading package Series'.

Out[10]=
        Cos[x]
```

That settles it! The derivative of the sine *is* the cosine. If you insist, we could make one more check. What does *Mathematica* say?

```
In[11]:=
        Sin'[x]

Out[11]=
        Cos[x]
```

The Derivative of the Cosine?

What is the derivative of the cosine? Most people guess the sine. But even though there is an appealing symmetry to such a rule, it can't be right. Look again at the plots of the sine and cosine shown on the preceding pages. The sine is positive when the cosine in falling. On that basis alone it cannot be the derivative of the cosine.

Let's try the idea that worked so well with the sine. Plot a difference quotient for the cosine and see if we can guess what curve it is.

```
In[12]:=
        Plot[
            (Cos[x + 0.01] - Cos[x])/0.01,
            {x,-2Pi,2Pi},
            PlotRange->{-1,1},
            Ticks->{{-2Pi,-Pi,0,Pi, 2Pi}, {-1,1}}
            ];
```

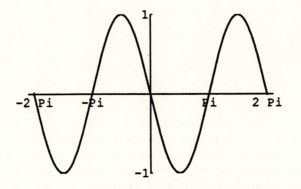

This curve has amplitude 1 and period 2π, but it is clearly neither the sine nor the cosine curve. It looks like a sine curve that has been flipped over the x-axis, $-\sin(x)$.

Check with a limit command.

```
In[13]:=
        Limit[(Cos[x + h] - Cos[x])/h, h -> 0]
```

```
Out[13]=
        -Sin[x]
```

There you have it: the derivative of the cosine is the negative of the sine.

The Derivative of Sin(2*x*)

What do you get when you differentiate sin(2*x*)? Many people guess cos(2*x*). But, as with the previous example, the instinctive first guess doesn't pan out. Seeing exactly why is a little tricky. For example, when you compare plots, cos(2*x*) looks like it might be a slope predictor for sin(2*x*).

```
In[14]:=
    Plot[
      Sin[2x], {x, 0, 2Pi},
      PlotLabel->"function, sin(2x)",
      Ticks->{{0,Pi,2Pi}, {-1,1}}
      ];
    Plot[
      Cos[2x], {x, 0, 2Pi},
      PlotLabel->"derivative?, cos(2x)",
      Ticks->{{0,Pi,2Pi}, {-1,1}}
      ];
```

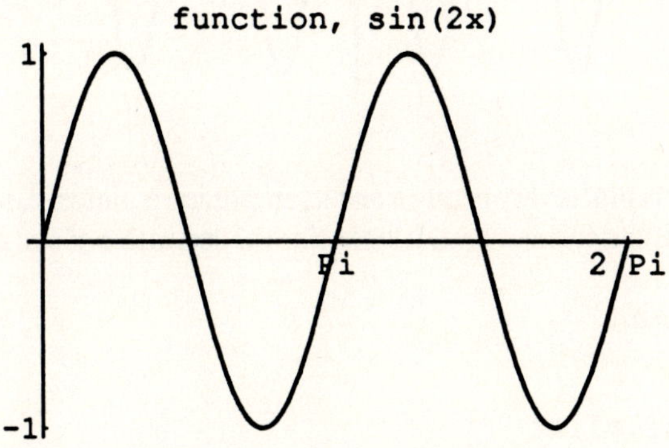

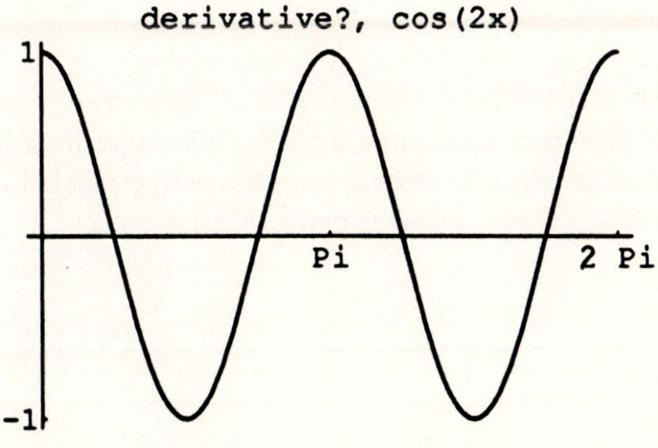

function, sin(2x)

derivative?, cos(2x)

Cos(2x) is positive when the sine graph is rising, negative when it is falling, and has zeros when the sine changes direction. It is however, not the derivative as a difference quotient approximation will quickly show.

```
In[15]:=
        Plot[(Sin[2(x + 0.01)] - Sin[2x])/0.01,
          {x, -2Pi,2Pi},
          PlotRange -> {-2.2,2.2},
          Ticks->{{-2Pi,-Pi,0,Pi,2Pi}, {-2,2}}
          ];
```

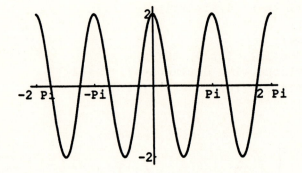

The period of this plot is very nearly π and its amplitude is almost 2, so the expression $2\cos(2x)$ describes it as a cosine curve. It looks like the derivative of $\sin(2x)$ is *not* $\cos(2x)$ but $2\cos(2x)$.

Ask *Mathematica*.

```
In[16]:=
        D[Sin[2x],x]

Out[16]=
        2 Cos[2 x]
```

The sequence of difference quotients did not lie. When you think about it, the form of the derivative does make sense. In order to complete two periods between 0 and 2π, as $\sin(2x)$ does, the functions needs to change twice as fast as $\sin(x)$.

Problems

1. Plot a difference quotient for the function $\sin(x/3)$ for small values of dx. This plot will approximate a cosine curve. What is its period and amplitude? Write out an expression for the curve in the form $a\cos(bx)$. Check your result by computing a derivative. Print out your plots and computations.

2. Plot a difference quotient for the function cos(5x). This plot will approximate a sine curve of the form *a* sin(*bx*). Which one is it? Print out your plots and computations.

3. The difference quotients for the function cos(x^2) have the form

$$\frac{Cos[(x + dx)^2] - Cos[x^2]}{dx}$$

Print out plots of the difference quotient and the function –sin(x^2) for *x* in the interval $-2\pi \le x \le 2\pi$. Use the plots to explain why the derivative of cos(x^2) is *not* –sin(x^2). (Be sure the 2's are *inside* the brackets.)

4. Here is a plot of the difference quotient for the tangent with $dx = 0.1$:

```
In[17]:=
          Plot[(Tan[x + 0.1]-Tan[x])/0.1,
          {x,-1.4,1.4},
          PlotRange->{0,5}
          ];
```

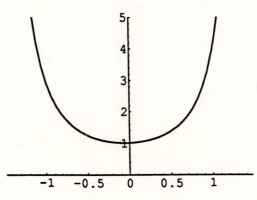

Plot the tangent function over the same *x*-interval. Print out both plots along with a paragraph explaining the relationship between them.

5. Try plotting the reciprocal of the difference quotient in Problem 4 for $-2\pi \le x \le 2\pi$. What is the period and amplitude of the curve? What is your guess for the derivative of the tangent?

13

Projectile Motion and Parametric Equations

Projectile Motion

A cannon is shot at an angle of 30 degrees above the horizontal with an initial velocity is 1000 feet per second. The motion of the cannon ball can be thought of as having two parts, horizontal motion and vertical motion. These two motions act together to produce the actual path of the projectile, which will look something like this:

The Projectile's Path

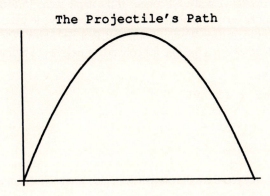

In order to describe this motion more precisely, assume that the vertical and horizontal axes in the picture above are the *x* and *y* axes and that the point of fire is the origin. As the projectile moves along the path of motion the *x* and *y* coordinates of its position change with time. Our aim here is to describe these coordinates as functions of time.

At the instant the cannon is fired, the cannon ball is going 1000 ft/sec at a 30 degree angle to the horizontal. The initial horizontal velocity, *xvel*(0), and vertical velocity, *yvel*(0), can be computed from the right triangle below.

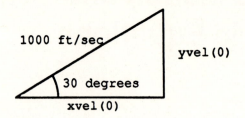

CONVERTING FROM DEGREES TO RADIANS

Some care must be exercised is computing the numerical values for these initial velocities because *Mathematica*'s trigonometric functions are programmed to read angles in radians, not degrees. So, for example, if you try to calculate the initial horizontal component of velocity, *xvel*(0), with the command `N[1000*Cos[30]]` you will get the wrong result. *Mathematica* will evaluate `Cos[30]` as the cosine of 30 *radians*.

There are a couple of ways of dealing with this problem. You could simply convert 30 degrees to $\pi/6$ radians in your head and use the command `N[1000*Cos[Pi/6]]` to get the initial horizontal velocity. Of course, if the angle were 35.14672 degrees the conversion would not be so easy, and you would probably have to get out a pencil and paper to figure out how to set it up. We don't want that. Pencils? Paper? Too old fashioned. So, try this: Teach *Mathematica* the rule for converting degrees to radians.

```
In[1]:=
        degrees = Pi/180
```

```
Out[1]=
        Pi
        ---
        180
```

Check to make sure you have the correct rule. You know for example that the sine of 30 degrees is 1/2.

```
In[2]:=
        N[Sin[30 degrees]]
```

```
Out[2]=
        0.5
```

Perfect. You would not have gotten the correct sine value if you had left off the word **degrees.**

```
In[3]:=
        N[Sin[30]]
```

```
Out[3]=
        -0.988032
```

That output is the sine of 30 radians or about 1700 degrees.

Note that although the command **N[Sin[30 degrees]]** looks as if we are asking *Mathematica* to compute in degrees, that is not the effect at all. The expression **30 degrees** actually just multiplies 30 by the conversion factor $\pi/180$, which returns the equivalent number of radians.

Now that we have the angle measure straightened out, ask for the initial horizontal velocity this way:

```
In[4]:=
        N[1000*Cos[30 degrees]]
```

```
Out[4]=
        866.025
```

Thus we know that the initial horizontal velocity is 866.025 ft/sec.

THE HORIZONTAL DISTANCE FORMULA

Since there are no forces acting on the projectile in the horizontal direction (we are ignoring air resistance), the speed in the horizontal direction is constant at 866.025 ft/sec as long as the projectile is aloft. Therefore, the horizontal distance x through which the projectile moves in t seconds is $x(t) = 866.025\ t$. Teach *Mathematica* the horizontal distance formula.

```
In[5]:=
        Clear[x,t];
        x[t_]:= 866.025*t
```

THE VERTICAL DISTANCE FORMULA

The force of gravity acts vertically, slowing the initial vertical velocity by 32.2 ft/sec for each second that the projectile is aloft. In other words, the vertical acceleration is −32.2 ft/sec². This means that the vertical velocity after t seconds will be the initial vertical velocity minus 32.2 times the time.

Here is the initial vertical velocity:

```
In[6]:=
        1000 * Sin[30 degrees]
```

```
Out[6]=
        500
```

After *t* seconds the vertical velocity will be

```
In[7]:=
        yvel = 500 - 32.2t
```

```
Out[7]=
        500 - 32.2 t
```

Antidifferentiation gives the formula for the vertical distance. Since $y(0) = 0$, the constant will be zero.

```
In[8]:=
        Clear[y,t]
        y[t_] := 500*t - 16.1*t^2
```

THE PATH OF THE PROJECTILE

Now, we can use the formulas for the projectile's horizontal and vertical distances to produce a table of positions. The table gives the location of the projectile at the end of each of its first 10 seconds aloft.

```
In[9]:=
        TableForm[
          Table[{t,x[t],y[t]}, {t,0,10}],
          TableHeadings->
            {None,
            {"t","x(t)\n\n", "y(t)\n\n"}
            }
          ]
```

```
Out[9]//TableForm=
         t      x(t)        y(t)

         0      0           0
         1      866.025     483.9
         2      1732.05     935.6
         3      2598.08     1355.1
         4      3464.1      1742.4
         5      4330.12     2097.5
         6      5196.15     2420.4
         7      6062.17     2711.1
         8      6928.2      2969.6
         9      7794.22     3195.9
        10      8660.25     3390.
```

The projectile moves to the right and up as time passes. For example, the second point in the table tells you that after 1 second, the projectile is at a position 866 feet to the right and 484 feet above the point at which it was fired.

Plot these positions.

```
In[10]:=
        ListPlot[
          Table[{x[t],y[t]}, {t, 0, 10}],
          PlotStyle->PointSize[.02]
          ];
```

The projectile rises and moves to the right simultaneously producing a curved path of motion. Let's see the rest of the path.

```
In[11]:=
        ListPlot[
          Table[{x[t],y[t]}, {t, 0, 20}],
          PlotStyle->PointSize[.02]
          ];
```

The t interval wasn't long enough to show the complete path. What should the t-range be in order to display the whole path? What value of t will make y zero again? (The x-axis tick marks are a mess, but we can deal with that later.)

```
In[12]:=
        Solve[y[t]==0, t]
```

```
Out[12]=
        {{t -> 31.0559}, {t -> 0}}
```

A little more than 31 seconds.

```
In[13]:=
        ListPlot[
          Table[{x[t],y[t]}, {t,0,31}],
          PlotStyle->PointSize[.02]
          ];
```

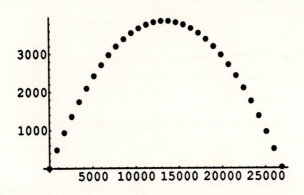

The horizontal ticks can be controlled by hand. Find out what x is at 31 seconds so we know the largest x-value.

```
In[14]:=
        x[31]
```

```
Out[14]=
        26846.8
```

Try this for the **Ticks** setting.

```
In[15]:=
        ListPlot[
          Table[{x[t],y[t]}, {t,0,31}],
          PlotStyle->PointSize[.02],
          Ticks->{{0,13000,26000}, Automatic}
          ];
```

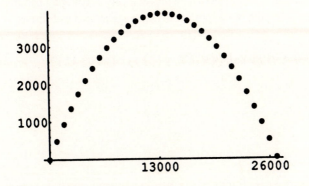

That's a little better. The projectile rises and falls and, after about 31 seconds, it lands 26,846.8 feet down range. From the spacing of the positions, which were calculated at 1-second intervals, we can see the vertical deceleration. The points bunch together at the top of the path where the vertical velocity is closest to zero. The maximum height will occur when the vertical velocity is exactly zero.

```
In[16]:=
        Solve[y'[t]==0, t]
```

```
Out[16]=
        {{t -> 15.528}}
```

So, when t = 15.528 seconds, the projectile will be at its highest position, 3881.94 feet above the ground.

```
In[17]:=
        y[15.5828]
```

```
Out[17]=
        3881.94
```

Parametric Curves

When a curve is given using two functions, such as the pair we used here to describe the path of the projectile, $x = 866.025\ t$ and $y = 500\ t - 16.1\ t^2$, the equations are called a *parametric* representation of the curve. The variable *t* is the *parameter*. We'll look at some additional examples of parametric curves in a minute, but first a couple of helpful details.

PARAMETRICPLOT

Mathematica has a built-in command which makes it very easy to plot parametric curves. For example, the whole path of the projectile can be generated as follows:

```
In[18]:=
        ParametricPlot[{x[t],y[t]}, {t,0,31}];
```

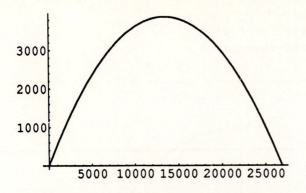

Practice Exercise 1: Plot the curve of motion for a projectile fired at an angle of 75 degrees with an initial velocity of 750 feet per second.

ANOTHER PARAMETRIC CURVE

This time we will begin with the a pair of equations which define the path of a moving object. Our task is to figure out what that motion is like. The input cell below defines the parametric equations and asks for a graph.

```
In[19]:=
       Clear[x,y,t];
       x[t_]:= 3Sin[t];
       y[t_]:= 3Cos[t];
       ParametricPlot[{x[t],y[t]}, {t,0,6}];
```

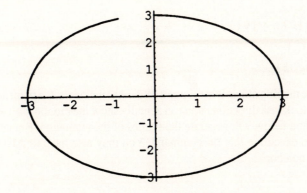

The shape is off because the vertical scale is different from the horizontal scale. The setting **AspectRatio->Automatic** tells *Mathematica* to use the same scale on both axes.

```
In[20]:=
        ParametricPlot[
            {x[t],y[t]}, {t,0,6},
            AspectRatio->Automatic
            ];
```

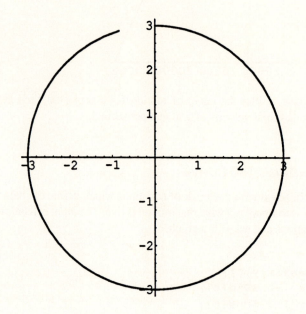

Clearly, the path is a circle of radius 3. The motion begins when $t = 0$.

```
In[21]:=
            {x[0], y[0]}
```

```
Out[21]=
            {0, 3}
```

Initially the moving object is at the point $(0, 3)$. As t increases, the point moves around the circle in a clockwise direction. You can actually show the motion with an animation. The **Do** command below will generate the frames of the animation. Remember to save your file before executing this **Do** command. You may also have to **Clear** the other graphs in your notebook.

```
In[22]:=
   Do[
     ParametricPlot[
       {x[t],y[t]}, {t,0,.5k},
       AspectRatio->Automatic,
       PlotRange->{{-3.5,3.5},{-3.5,3.5}}
       ],
     {k,1,12}
     ];
```

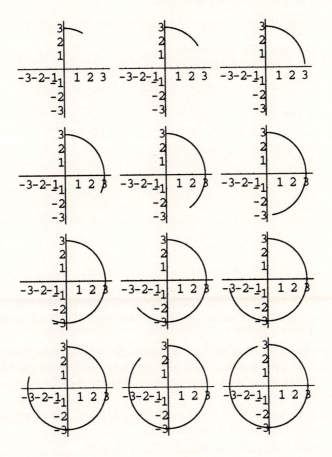

When the plots are on your screen, select the bracket which contains them all and then select **Animate Selected Graphics** from the **Graph** menu. Remember that the speed of the animations can be controlled by the buttons in the lower left corner of the window. For more detailed instructions on controlling animations, see Exploration Five.

The animation clearly shows the motion beginning at (0, 3) and moving clockwise around the circle.

Practice Exercise 2: Graph the parametric curve $x = 9\cos(t) - \cos(9t)$, $y = 9\sin(t) - \sin(9t)$ through one period of the sine.

Code for the Practice Exercises

1. ```
ParametricPlot[
    { (750 Cos[75 degrees])t,
      (750 Sin[75 degrees])t - 16.1t^2 },
    {t,0,45}
    ];
```

2. ```
ParametricPlot[
    {9 Cos[t]-Cos[9t], 9 Sin[t]-Sin[9t]},
    {t, 0, 2Pi}
    ];
```

Problems

1. Parametric equations of the form $x(t) = a + b\cos(t)$ and $y(t) = c + d\sin(t)$, $0 \le t \le 2\pi$, define curves in the xy-plane. Make parametric plots for several values of a and c using $b = 2$ and $d = 4$. Write a paragraph explaining how changes in the values of a and c affect the shape and position of the curves. Illustrate your remarks with plots. (Hint: Set **AspectRatio** to **Automatic**.)

2. Referring to Problem 1, fix values for a and c and make parametric plots for several values of b and d. Write a paragraph explaining how changes in the values of b and d affect the shape and position of the curves. Illustrate your remarks with plots. (Hint: Set **AspectRatio** to **Automatic**.)

3. Find a parametric representation for the curves described below. Turn in the commands and a plot that shows you have the correct curve.

   (a) The left half of the circle with radius 5 and center at (4,3).

   (b) The top half of the ellipse with center at (3,1), major axis parallel to the x-axis, length of major axis 10 and length of the minor axis 4.

4. Plot the parametric curve $x = \cos(4t) \cos(t)$, $y = \cos(4t) \sin(t)$, $0 \le t \le 2\pi$, with the initial point enlarged. Assume that the curve is the path of a moving object and mark arrows on the plot by hand to indicate the direction of motion. (Hint: A **Do** command can help you to see the direction of motion.)

5. At what angle should a cannon on the ground be fired to reach the greatest possible horizontal distance? Assume that the initial velocity is 1000 ft/sec. You may do this by solving appropriate equations or by experimenting with different angles. Include a plot illustrating your solution. Also, find the maximum height the cannon ball attains when this angle of fire is used.

14

Area Predicting Formulas

We are interested in the general problem of calculating the area under a curve $y = f(x)$ over an interval on the x-axis, $a \leq x \leq b$.

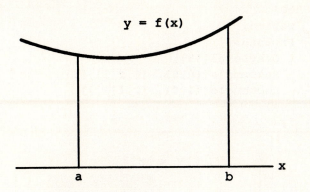

Begin with the following specific problem:

> Find the area under the curve $y = 3x^2$, for $0 \leq x \leq 2$.

Type in and enter the following *Mathematica* commands, which will draw the region under the curve.

```
In[1]:=
      f[x_]:= 3x^2;
      region = Plot[
          f[x], {x,0,2},
          Epilog->{Line[{{2,0}, {2,12}}]},
          AxesLabel->{" x","y"},
          PlotLabel->"y = 3x^2"
          ];
```

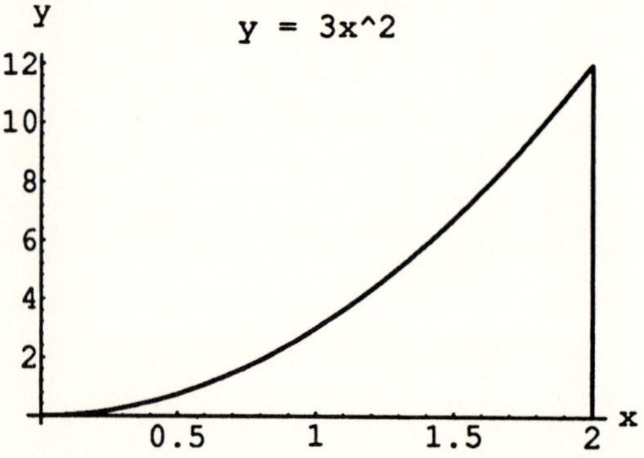

Our goal is to determine the area of this region as exactly as possible. We can begin by getting a rough approximation deduced from the following picture.

```
In[2]:=
      Show[
        region,
        Prolog->{{
          GrayLevel[.8],
          Rectangle[{0,0}, {1,f[1]}],
          Rectangle[{1,0}, {2,f[2]}]
          }}
        ];
```

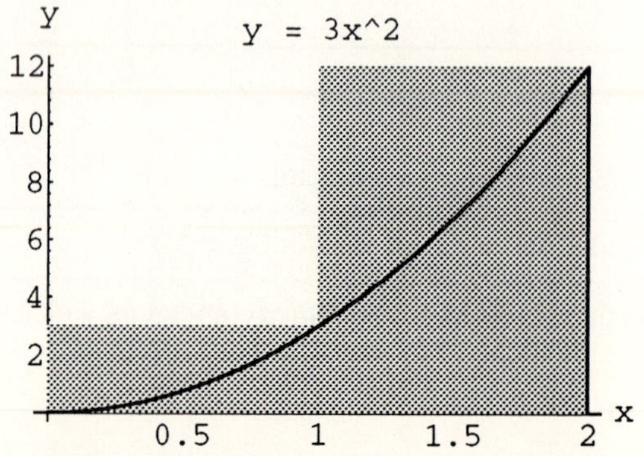

Note: The `Rectangle` command is a new one. It takes two points as its argument and uses the first point as the lower left corner of the rectangle and the second point as the upper right corner.

Look at the two rectangles above. The idea is this: We will add together their areas to get an approximation to the area of the region. Of course, the rectangles extend beyond the region, so we know that the total will be larger than the area we are looking for, but it will provide an upper bound for the area. Once we have this upper bound, the next step will be the construction of a lower bound in a similar way using rectangles. When we have the area bounded, we still won't know exactly what the area is, but we will have an idea of about how big it is. At that point we can worry about how to make the upper and lower bounds more precise.

Look at the two rectangles carefully: Each has a base that is one unit long, but their heights are different. The height of the rectangle on the left, the shorter one, is the vertical distance from the point $(1, 0)$ on the x-axis to the curve, and so it is equal to $f(1)$ or 3. The height of the rectangle on the right is $f(2)$ or 12. The total area in the two rectangles is $1*3 + 1*12 = 15$ square units. Of course, *Mathematica* can do this calculation in a flash.

The area in these two rectangles is

```
In[3]:=
        1*f[1] + 1*f[2]
```

```
Out[3]=
        15
```

Here's a picture which gives a lower bound:

```
In[4]:=
        Show[
          region,
          Prolog->{{GrayLevel[.8],
          Rectangle[{1,0},{2,f[1]}]}}
          ];
```

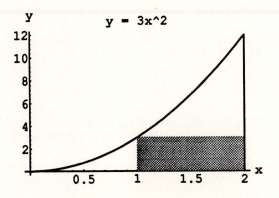

The area of the rectangle is $1*f(1) = 3$.

Thus we know the area of the region is between 3 and 15 square units, $3 \leq A \leq 15$. Now the task is to narrow the interval between the upper and lower bounds by calculating more precise estimates. A cinch. All we have to do is use more rectangles.

Here is the picture for a upper bound based on four rectangles:

```
In[5]:=
        Show[
          region,
          Prolog->{{
            GrayLevel[.8],
            Rectangle[{0,0},{.5,f[.5]}],
            Rectangle[{.5,0},{1,f[1]}],
            Rectangle[{1,0},{1.5,f[1.5]}],
            Rectangle[{1.5,0},{2,f[2]}]
            }}
          ];
```

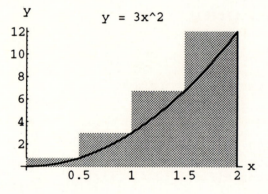

The area in these four rectangles is

```
In[6]:=
        0.5*f[.5] + 0.5*f[1] + 0.5*f[1.5] + 0.5*f[2]

Out[6]=
        11.25
```

A little terminology: The four rectangles above are called *circumscribed* rectangles.

Let's calculate the corresponding lower bound. Here's the picture. It will show four *inscribed* rectangles.

```
In[7]:=
    Show[
      region,
      Prolog->{{
        GrayLevel[.8],
        Rectangle[{0,0},{.5,f[0]}],
        Rectangle[{.5,0},{1,f[.5]}],
        Rectangle[{1,0},{1.5,f[1]}],
        Rectangle[{1.5,0},{2,f[1.5]}]
        }}
      ];
```

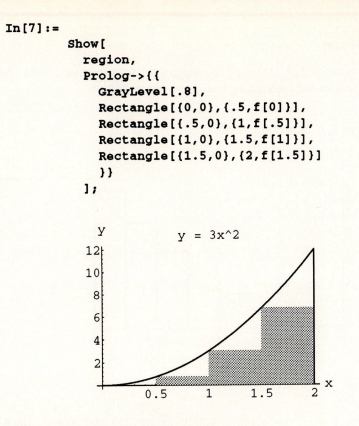

The sum below gives the total area in the rectangles.

```
In[8]:=
    0.5*f[0] + 0.5*f[.5] + 0.5*f[1] + 0.5*f[1.5]

Out[8]=
    5.25
```

We're closing in (slowly) on the value of A: $5.25 \leq A \leq 11.25$.

Using Area.m

Obviously, we can obtain closer and closer approximations to the area by using more rectangles. We're going to do eight rectangles next. Don't groan. The package **Area.m** contains commands which will do most of the work for you. But first you will have to find and load the package which is on your *Exploring Calculus* disk in the **Packages** file. (See Exploration Five for detailed instructions on loading a package.)

```
In[9]:=
        <<Area.m
```

Once the package is loaded, type in the next command. It will draw the region with eight circumscribed rectangles and calculate the total area in the rectangles.

```
In[10]:=
        AreaR[f,0,2,8]
```

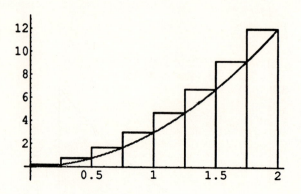

```
Out[10]=
        9.5625
```

Isn't that a time saver? A word about the name of the function used: **AreaR**. The **R** means that the height of each of the the rectangles is drawn over the right-hand endpoint of its base. Since the curve is increasing over the interval from 0 to 2, the rectangles circumscribe the region, and so their total area gives an upper bound for the area of the region.

The lower bound based on eight rectangles is just as easy to produce.

```
In[11]:=
        AreaL[f,0,2,8]
```

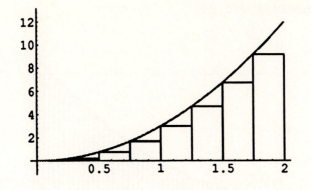

Out [11] =
        6.5625

Now the bounds on area are 6.5625 < *A* < 9.5625.

Here is an approximation based on rectangles drawn over the midpoint of the base of the rectangles.

In[12]:=
        AreaMP[f,0,2,8]

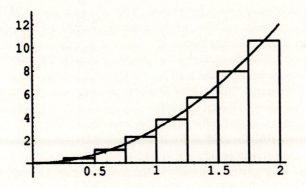

Out [12] =
        7.96875

You can see from the picture that the total area in these rectangles is closer to the area of the region than either the upper or the lower bounds based on eight rectangles. Each rectangle in the midpoint approximation extends beyond the region, but each rectangle also misses part of the region so there is a cancelling out of the error.

Since *Mathematica* is doing all of the work, let's look at some pictures with lots of rectangles. Here are 50 circumscribed rectangles:

```
In[13]:=
        AreaR[f,0,2,50]
```

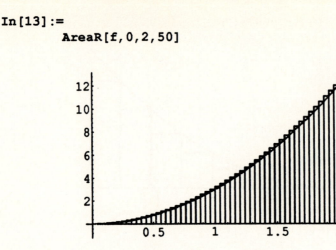

```
Out[13]=
        8.2416
```

Practice Exercise 1: Find the total area using 50 inscribed rectangles and also using 50 midpoint rectangles. (*Ans.* 7.7616, 7.9992)

The upper and lower bounds for *A* based on 50 rectangles give: 7.7616 < *A* < 8.2416. The result of the midpoint calculation finds a number, 7.9992, which is closer to the true area than either of the bounding numbers.

You can see that as the number of rectangles increases, they provide a better and better fit to the region and, so the total area in the rectangles gives a better and better approximation to the area. This means that we can use the **Area.m** functions to calculate more precise estimates for the area by increasing the number of rectangles. You notice, though, that it takes a while for the pictures to be produced when lots of rectangles are asked for. So, how's this for an idea: We'll skip the pictures and just ask for the numerical total of the rectangular areas. Let's also take advantage of the fact that the midpoint calculations are giving better estimates to the area than either the upper or the lower bounds.

When you wish only the numerical value of the total area in the approximating rectangles and don't want to see the picture, enter an N command.

```
In[14]:=
        NAreaMP[f,0,2,50]

Out[14]=
        7.9992
```

Here's a table of midpoint calculations for 50, 60, 70, 80, 90, and 100 rectangles. It may take your machine a while to produce the table. Be patient.

```
In[15]:=
        TableForm[
          Table[
            {n, NAreaMP[f,0,2,n]},
            {n,50,100,10}
            ],
          TableHeadings->
            {None,
            {"number of\nrectangles\n",
            "approx. area\nunder 3x^2\n"
              <>"0 ≤ x ≤ 2\n"
            }}
          ]
```

```
Out[15]//TableForm=
          number of      approx. area
          rectangles     under 3x^2
                         0 ≤ x ≤ 2

          50             7.9992
          60             7.99944
          70             7.99959
          80             7.99969
          90             7.99975
          100            7.9998
```

We'd guess that exact area of the region is 8 square units. Test the guess with 1000 rectangles. (A word of warning: This calculation takes several minutes even with a fast Macintosh, so if your machine is slow, you might want to take our word for the output.) The $N[...,15]$ will force *Mathematica* to give extra digits in the answer.

```
In[16]:=
        N[NAreaMP[f,0,2,1000],15]
```

```
Out[16]=
        7.999998
```

Close enough. We declare the area to be eight square units!

Move on to the next area calculation. It will go a lot faster, since we have a pretty good idea of how to approach the problem.

We'll try a different region under the same curve, $y = 3x^2$. This time use the interval $0 \le x \le 3$. Start with a relatively small number of rectangles so you get results quickly.

In[17]:=
      AreaMP[f,0,3,10]

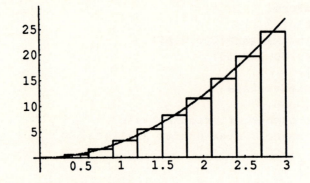

Out[17]=
      26.9325

Generate a short table.

In[18]:=
```
    TableForm[
      Table[
        {n,NAreaMP[f,0,3,n]},
        {n,10,50,10}
        ],
      TableHeadings->
        {None,
        {"number of\nrectangles\n",
        "approx. area\nunder 3x^2,\n"
          <>"0 ≤ x ≤ 3\n"
        }}
        ]
```

Out[18]//TableForm=

number of rectangles	approx. area under 3x^2, 0 ≤ x ≤ 3
10	26.9325
20	26.9831
30	26.9925
40	26.9958
50	26.9973

It looks like the area here is 27 square units. Test the guess with a large number of rectangles.

In[19]:=
        AreaMP[f,0,3,100]

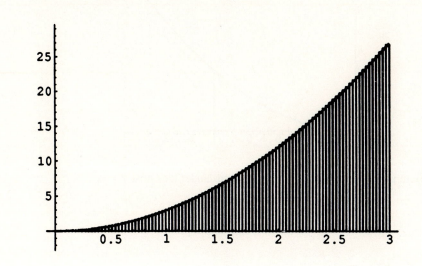

Out[19]=
        26.9993

It certainly looks very close to 27 square units.

Practice Exercise 2: What is the area of this region?

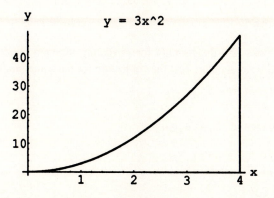

Approximate it with a large number of rectangles.

Practice Exercise 3: How about this region? Use a `Table` command with an increasing number of rectangles.

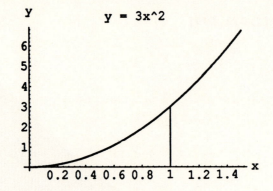

Here is a summary of the results for areas under the curve $y = 3x^2$.

```
    Interval          Area Under
                      y = 3x^2

    0 ≤ x ≤ 1             1
    0 ≤ x ≤ 2             8
    0 ≤ x ≤ 3            27
    0 ≤ x ≤ 4            64
```

Based on the information in the table what number would you guess for the area under the curve between $x = 0$ and $x = 5$? Almost certainly 125 sq. units. Let's see.

```
In[22]:=
        NAreaMP[f,0,5,50]
```

```
Out[22]=
        124.987
```

It certainly looks like the formula for predicting the area under this curve and above the interval $0 \le x \le t$ is t^3. Test the conjecture for the some random interval, say $0 \le x \le 5.6$

```
In[23]:=
        NAreaMP[f,0,5.6,10]
```

```
Out[23]=
        175.177
```

Is this close to $(5.6)^3$?

```
In[24]:=
        5.6^3

Out[24]=
        175.616
```

Not bad.

Code for the Practice Exercises

1. `AreaL[f,0,2,50]`, `AreaMP[f,0,2,50]`

2. `NAreaMP[f,0,4,50]`

3. ```
   TableForm[
     Table[
       {n,NAreaMP[f,0,1,n]}, {n,10,50,10}
       ],
     TableHeadings->
       {None,
       {"number of\nrectangles\n",
       "approx. area\nunder 3x^2,\n"
         <>"0 ≤ x ≤ 1\n"
       }}
     ]
   ```

Problems

1. Use the area predicting formula derived above to calculate the area under the curve $y = 3x^2$ for $0 \le x \le 10$, and check the result using **NAreaMP**.

2. Use the area predicting formula derived above to calculate the area under the curve $y = 3x^2$ for $2 \le x \le 10$.
   ```

3. Fill in the table below for approximate areas under the curve $y = 4x^3$. Use the function **NAreaMP**. Print out a picture of the rectangles in each case when the number of rectangles is 30.

Interval	Number of Rectangles	Area in the Rectangles
$0 \le x \le 1$	5	
$0 \le x \le 1$	10	
$0 \le x \le 1$	20	
$0 \le x \le 1$	30	
$0 \le x \le 2$	5	
$0 \le x \le 2$	10	
$0 \le x \le 2$	20	
$0 \le x \le 2$	30	
$0 \le x \le 3$	5	
$0 \le x \le 3$	10	
$0 \le x \le 3$	20	
$0 \le x \le 3$	30	

What do you guess is the area predicting formula for the curve $y = 4x^3$? Test your guess for a few random intervals.

4. Repeat the exercise above for the curve $y = 2x$. Prove, using geometry, that your guess at an area predicting formula is correct.

5. Find an upper and lower bound for the area that is under the curve $y = \sin(x)$, and above the interval $0 \le x \le \pi/2$. Use 20 rectangles. Print out a graph of the approximating rectangles and their total area.

15

Area Between Curves

Consider the problem of computing the area caught between the two curves
$y = x^4 - 2x$ and $y = 2x^2$.

The region can be pictured with a plot command.

```
In[1]:=
        f[x_]:= x^4 - 2x + 3
        g[x_]:= 2x^2 + 1
        Plot[{f[x],g[x]},{x,-2,2}];
```

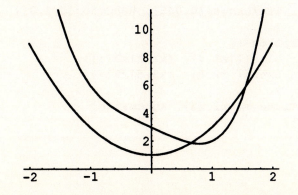

The plot shows the curves intersecting twice at either end of a small region. We need to find out if there are any intersections other than the two currently visible. If there are not, then the region we see between the curves is the region whose area we need. Use **Solve** to find the *x*-coordinates of the intersections.

```
In[2]:=
        N[Solve[f[x]==g[x], x]]
```

```
Out[2]=
        {{x -> 1.56917}, {x -> 0.659584},
         {x -> -1.11438 + 0.830976 I},
         {x -> -1.11438 - 0.830976 I}}
```

The two real solutions are the ones we see in the plot. The complex solutions have no meaning for our current problem. Since we will need these two real numbers to solve our problem, let's assign them variable names so it will be easy to refer to them later.

```
In[3]:=
        x1 = 0.659584
        x2 = 1.56917
```

```
Out[3]=
        1.56917
```

Edit the plot command to get a better view of region between the two curves, with the x_1 and x_2 values marked on the horizontal axis and lines drawn from them to the curves. The **PlotStyle** setting cause the graph of $f(x)$ to be drawn thickly in red and $g(x)$ more thinly in green. If you do not have a color monitor, omit the **RGBColor** parts of the setting.

```
In[4]:=
        Plot[
          {f[x],g[x]}, {x,-.5,2},
          PlotStyle->{
            {Thickness[0.01],  RGBColor[1,0,0]},
            {Thickness[0.005], RGBColor[0,1,0]}
            },
          Epilog->{
            Line[{{x1,0}, {x1,f[x1]}}],
            Line[{{x2,0}, {x2,f[x2]}}]
            },
          Ticks->{{x1,x2}, Automatic}
          ];
```

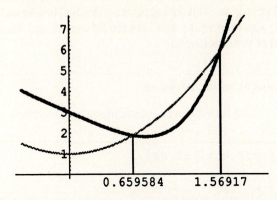

Clearly the area between the curves will be the area under the top curve, $g(x)$, for the interval $0.659584 \leq x \leq 1.56917$ minus the area under the bottom curve, $f(x)$, over exactly the same interval. The command below calculates the area under the top curve.

```
In[5]:=
      Integrate[g[x], {x,x1,x2}]

Out[5]=
      3.29412
```

For the bottom curve, you use

```
In[6]:=
      Integrate[f[x], {x,x1,x2}]

Out[6]=
      2.57929
```

Subtracting the two integrals gives the area between the two curves.

```
In[7]:=
      Integrate[g[x], {x,x1,x2}] -
        Integrate[f[x], {x,x1,x2}]

Out[7]=
      0.714834
```

Good. The area between the curves is about seven-tenths of a square unit.
    You can also get the same result with a single integration.

```
In[8]:=
      Integrate[g[x]-f[x], {x,x1,x2}]

Out[8]=
      0.714834
```

Let's check to see if this result is reasonable. Begin by loading the package **AreaBetweenCurves.m**. This package contains the function **AreaBet**, which will draw rectangles to approximate the area between the curves, and **NAreaBet**, which gives the total area of those rectangles..

In[9]:=

    <<AreaBetweenCurves.m

Now, ask for the area between to be approximated by 20 rectangles.

In[10]:=

    AreaBet[f,g,x1,x2,20]

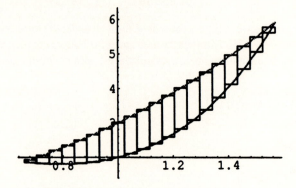

Get the numerical total of the areas of these 20 rectangles and see if it is close to 0.714834, the value we got using **Integrate**.

In[11]:=

    NAreaBet[f,g,x1,x2,20]

Out[11]=

    0.715754

It looks like our **Integrate** calculation was right on the money.

In[12]:=

    Clear[f,g,x1,x2]

In the prior example, we saw that the area between the curves could be found in a single integration using an integrand of $f(x) - g(x)$, a nice shortcut. But beware; it doesn't always work.

> Find the area caught between the curves
> $$y = -28 + 49x - 12x^2 + x^3$$
> and $y = 10x.$

Begin by determining where the curves intersect.

```
In[13]:=
        Clear[g,f];
        f[x_]:= -28 + 49*x - 12*x^2 + x^3;
        g[x_]:= 10*x;
        N[Solve[f[x] == g[x], x]]

Out[13]=
        {{x -> 7.}, {x -> 4.}, {x -> 1.}}
```

Set up a plot command to show the curves and their intersection points.

```
In[14]:=
        Plot[{f[x],g[x]}, {x,0,8}];
```

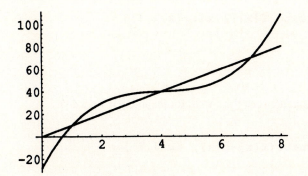

Let's get a rough idea of what the area is.

```
In[15]:=
        AreaBet[f,g,1,7,10];
        NAreaBet[f,g,1,7,10]
```

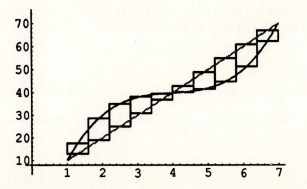

Out[15]=
        41.31

About 41 square units. Try integrating $f(x) - g(x)$ between $x = 1$ and $x = 7$. The output won't be even close to the actual area value.

In[16]:=
        Integrate[f[x]-g[x], {x,1,7}]

Out[16]=
        0

Is this reasonable? No. What's the difficulty? The problem boils down to this: The curves cross above $x = 4$. Between $x = 1$ and $x = 4$, f is the top curve and g the bottom curve, so in this interval the integrand $f(x) - g(x)$ is appropriate. But for $4 \leq x \leq 7$, g becomes the top curve, so the integrand over the second interval should be $g(x) - f(x)$.

In[17]:=
        Integrate[f[x]-g[x], {x,1,4}]

Out[17]=
        $\frac{81}{4}$

This is the correct area between the curves for the interval $1 \leq x \leq 4$.

In[18]:=
        Integrate[f[x]-g[x], {x,4,7}]

Out[18]=
        $-\left(\frac{81}{4}\right)$

This result is negative and so can't be the area for the second part of the region. The next integral computation will return the total area between the two curves.

In[19]:=
        Integrate[f[x]-g[x], {x,1,4}] +
          Integrate[g[x]-f[x], {x,4,7}]

Out[19]=
        $\frac{81}{2}$

```
In[20]:=
        N[%]

Out[20]=
        40.5
```

This result squares with the approximation.

Problems

1. Find the area caught between the curves $y = x^2$ and $y = 16$.

2. Find the area caught between the curves $y = x^3 - 12x$ and $y = 5x$.

   The command `FindRoot[f[x]==g[x], {x, a}]` finds one solution to the equation `f[x] == g[x]`, the one which is near the number a. You will find it helpful in determining the intersection points for the curves in Problems 3 and 4. The `Solve` command only works when f and g are polynomials. `FindRoot` will work for any functions, but it only gives one root at a time, you need to specify a value of **a** that is close to the root you want. So if there are three intersections, you need to use `FindRoot` three times. You can pick a value for **a** by estimating the value of the root from a plot.

3. Find the area caught between the curves $y = 2^x$ and $y = x^2$. Use `Integrate` to find the area, and use `AreaBet` to test the reasonableness of your result.

4. Find the area caught between the curves $y = x^2$ and $y = x^2 \sin(x)$, above the interval $-4\pi \le x \le 4\pi$.

16

Average Value of a Continuous Function

> A point is moving along the coordinate line in such a way that after t seconds
> its velocity, in feet per second, is given by the formula
>
> $$v(t) = 3t^2.$$
>
> Find the *average velocity* during the time interval $5 \le t \le 8$.

Method One: Using Integration

One way to get the average velocity is to take the total distance travelled and divide by
the time. For the given problem we can find the distance by integrating the velocity from
$t = 5$ to $t = 8$.

```
In[1]:=
      Clear[v,t];
      v[t_]:= 3t^2;
      Integrate[v[t],{t,5,8}]

Out[1]=
      387
```

So, in the 3-second time interval the point moves 387 feet, for an average velocity of 387/3 = 129 feet per second.

Look at a plot of the velocity together with the average velocity:

```
In[2]:=
        Plot[{v[t], 129}, {t,5,8}];
```

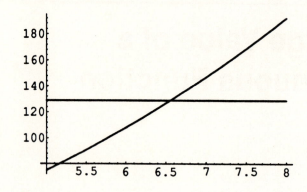

The horizontal line does look like a reasonable "average" for this velocity in the interval from $t = 5$ to $t = 8$. It splits the velocity values about in the middle.

The idea here can easily be extended to other functions, even if they do not represent velocity. To find the average value of $f(x)$ on the interval $a \le x \le b$, integrate f from a to b, and divide by $(b - a)$.

On the other hand, a person who knew nothing about integration, but had *Mathematica*'s power at hand, might go about the computation of the average value for a function in a very different way. This person might say, "Hey, I know how to average a list of numbers. All you do is count the numbers in the list, then add them up and divide by the count. So, I'll figure out the average velocity by totaling the velocity values and dividing by the number of values I summed."

The interesting thing is that this method can be made to work quite well, even though it seems at first glance that it could not possibly lead to anything meaningful. After all, the point speeds up continuously, and so it has a different velocity at each instant in time between $t = 5$ and $t = 8$. Consequently, summing the velocities and dividing by the number of them seems to give the dead-end ratio ∞/∞. Method Two shows you how you can avoid the dead end and apply the ordinary notion of average to a continuously changing quantity.

Method Two: Using Summation

Start small. Let's see how to get *Mathematica* to average a finite list of velocity values. Here is a list of six values:

```
In[3]:=
        Clear[v,t];
        v[t_]:= 3t^2;
        TableForm[
           Table[{t,v[t]}, {t,5.5,8,.5}],
           TableHeadings->{None, {"t\n","v(t)\n"}}
           ]
```

```
Out[3]//TableForm=
              t        v(t)

              5.5     90.75
              6.      108.
              6.5     126.75
              7.      147.
              7.5     168.75
              8.      192.
```

Add the six velocities.

```
In[4]:=
        Sum[v[t], {t,5.5,8,0.5}]
```

```
Out[4]=
        833.25
```

Look carefully at the **Sum** command, which is introduced here. Notice that the iterator {t, 5.5, 8, 0.5} is the same one used in the **Table** command, and it has the same effect. It causes velocity values to be computed for times beginning at $t = 5.5$ seconds, ending at $t = 8$ seconds, with increments of 0.5 seconds. The way in which the commands use the six values is quite different. **Table** puts them in a list and displays the list. **Sum** adds them up and displays the total.

Divide the total by 6 in order to average the six velocities.

```
In[5]:=
        %/6
```

```
Out[5]=
        138.875
```

This result, 138.875 ft/sec, is not the true average of the velocity over the interval, it is just the average of the six values we chose. But we would expect it to be close to the true average, and since we already know the correct average is 129, we see it is not too far off. The average of the six points is too large, since the moving point is speeding up and the table values were all computed at the *end* of half-second intervals.

By finding the average of six values at the beginning of the half-second time intervals we can get an approximate average that will be too small.

```
In[6]:=
        Sum[v[t], {t,5,7.5,0.5}]/6
```

```
Out[6]=
        119.375
```

These two finite averages establish a range in which the true average must lie.

$$119.375 < \text{true average} < 138.875$$

The range can be narrowed by using more velocity values spaced at shorter time intervals. How about a quarter of a second? Here is the table of velocity values computed at the end of the quarter-second intervals:

```
In[7]:=
        Clear[v,t];
        v[t_]:= 3t^2;
        TableForm[
          Table[{t,v[t]}, {t,5.25,8,.25}],
          TableHeadings->{None, {"t\n","v(t)\n"}}
          ]
```

```
Out[7]//TableForm=
              t         v(t)

            5.25      82.6875
            5.5       90.75
            5.75      99.1875
            6.        108.
            6.25      117.188
            6.5       126.75
            6.75      136.687
            7.        147.
            7.25      157.687
            7.5       168.75
            7.75      180.187
            8.        192.
```

Here is the new "too large" average velocity based on quarter-second intervals:

```
In[8]:=
        Sum[v[t], {t,5.25,8,0.25}]/12
```

`Out[8]=`
>   133.906

This is a table of velocities computed at the beginning of the quarter-second time intervals:

`In[9]:=`

```
TableForm[
  Table[{t,v[t]}, {t,5.0,7.75,0.25}],
  TableHeadings->{None, {"t\n","v(t)\n"}}
  ]
```

`Out[9]//TableForm=`

t	v(t)
5.	75.
5.25	82.6875
5.5	90.75
5.75	99.1875
6.	108.
6.25	117.188
6.5	126.75
6.75	136.687
7.	147.
7.25	157.687
7.5	168.75
7.75	180.187

Here is the average of the right-hand column:

`In[10]:=`

```
Sum[v[t], {t,5.0,7.75,0.25}]/12
```

`Out[10]=`
>   124.156

We now have

$$124.156 < \text{true average} < 133.906$$

These numbers, 124.156 and 133.906, are closer to the true average because more velocities were used in calculating them. To get very close we need to sum a very large number of velocities with a very small lapsed time between them. The more velocities we use in the sum and the shorter the time interval between them the closer we will be to the true average.

Suppose we decide to use 50 velocity values calculated at equal time intervals. between 5 seconds and 8 seconds. The time interval between calculations will be $(8 - 5)/50 = 0.06$. The **Sum** command which totals the 50 velocities at the end of each time interval is

```
In[11]:=
        Sum[v[t], {t,5.06,8,0.06}]
```

```
Out[11]=
        6508.59
```

Divide by 50 for the average.

```
In[12]:=
        %/50
```

```
Out[12]=
        130.172
```

Practice Exercise 1: Calculate the average of the 50 velocities at the beginning of the time intervals. (*Ans.* 127.832 ft/sec)

Practice Exercise 2: Calculate the averages of 100 velocities computed at the beginning and the end of equally spaced times. (*Ans.* 128.805 ft/sec and 129.585 ft/sec)

Here are the averages based on 1000 velocity calculations.

```
In[13]:=
        Sum[v[t], {t,5.003,8,0.003}]/1000
```

```
Out[13]=
        129.059
```

```
In[14]:=
        Sum[v[t], {t,5.,7.997,0.003}]/1000
```

```
Out[14]=
        128.942
```

The table below summarizes the calculations we've made so far. (We included high and low entries for 10,000 velocities, which took a fairly long time to compute.)

# of vels	time interval	average velocity low	average velocity high
6	0.5	119.375	138.875
12	0.25	124.156	133.906
50	0.06	127.832	130.172
100	0.03	128.8	129.585
1000	0.003	128.942	129.059
10000	0.0003	128.994	129.006

It appears that the low and high estimates are converging to 129 ft/sec, exactly the same value which the integral calculation of Method One gave us. This suggests that there must be some connection between the limiting value of these finite averages and the integral of the velocity function. Let's try to understand that connection.

Take a look at the finite average computation in a general situation. We choose n points $t_1, t_2, \ldots, t_n$, equally spaced in the interval $5 \leq x \leq 8$, and add the function values at those points. If we set $dt = (8 - 5)/n$, the points can be written $t_k = 5 + k*dt$.

In[15]:=
```
        Clear[v,t,k,dt];
        t[k_]:= 5+k*dt
```

Take $n = 10$ for an illustration. Then the average calculation begins with

In[16]:=
```
        Sum[v[t[k]], {k,1,10}]
```

Out[16]=
```
        v[5 + dt] + v[5 + 2 dt] + v[5 + 3 dt] +
          v[5 + 4 dt] + v[5 + 5 dt] + v[5 + 6 dt] +
          v[5 + 7 dt] + v[5 + 8 dt] + v[5 + 9 dt] +
          v[5 + 10 dt]
```

This is the sum for the average, but it is very similar to the sums used in defining an integral. If we multiply through by dt it becomes exactly like one of those sums.

In[17]:=
```
        Expand[%*dt]
```

Out[17]=
```
        dt v[5 + dt] + dt v[5 + 2 dt] +
          dt v[5 + 3 dt] + dt v[5 + 4 dt] +
          dt v[5 + 5 dt] + dt v[5 + 6 dt] +
          dt v[5 + 7 dt] + dt v[5 + 8 dt] +
          dt v[5 + 9 dt] + dt v[5 + 10 dt]
```

If the value of n is large this expression will be very nearly the area under the velocity curve between $t = 5$ and $t = 8$. Here is the picture for $n = 10$.

Learning Resources
Centre

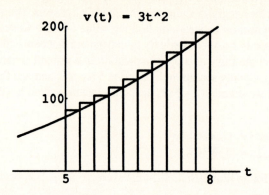

As n becomes large and dt small, the limit of this sum is the integral of $v(t)$ over the interval $5 \le t \le 8$. Consequently, the following equation is very nearly true, and the larger n is the truer it is.

```
Sum for the Average * dt = The integral
```

Now, divide both sides by $8 - 5$. Why? Because then the quotient on the right will be the integral expression for average velocity.

$$\frac{\texttt{Sum for the Average * dt}}{\texttt{8 - 5}} = \frac{\texttt{The integral}}{\texttt{8 - 5}}$$

But $8 - 5 = n\ dt$

$$\frac{\texttt{Sum for the Average * dt}}{\texttt{n dt}} = \frac{\texttt{The integral}}{\texttt{8 - 5}}$$

Cancel **dt**. The left-hand fraction is just the ordinary average of a very large number of velocities.

$$\frac{\texttt{Sum for the Average}}{\texttt{n}} = \frac{\texttt{The integral}}{\texttt{8 - 5}}$$

We conclude that as you use more and more velocities, the average calculated by adding the velocities at those points and dividing by the number of points approaches the integral over the interval divided by the length of the interval. This explains why methods one and two yielded the same result. And it further explains why mathematicians have decided on the following definition for the average value of a continuous function over an interval from a to b.

The *average value* of $f(x)$, on the interval $a \leq x \leq b$, is

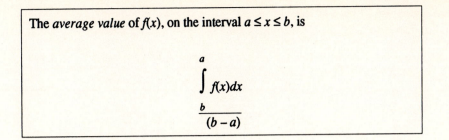

$$\frac{\int_b^a f(x)dx}{(b-a)}$$

Code for the Practice Exercises

1. `Sum[v[t], {t,5.0,7.94,0.06}]/50`
2. `Sum[v[t], {t,5.0,7.97,0.03}]/100`
   `Sum[v[t], {t,5.01,8.0,0.03}]/100`

Problems

1. Calculate the average velocity of a point with velocity $v = 3t^2$ for the interval $2 \leq t \leq 7$ using

   (a) a finite average of 100 velocities;
   (b) an integration.

2. On the average, what will be the square root of all real numbers between 1 and 10? First estimate the average using 50 square roots, then find the average with an integration.

3. In Exploration Eight we investigated the motion of a rocket whose height function was given by the two-part rule:

   $h_1(t) = 40t^2, 0 \leq t \leq 60$
   $h_2(t) = 144{,}000 + 4800(t - 60) - 16(t - 60)^2, \ 60 < t \leq 210$

   (a) Find the rocket's average height above the ground in the time interval from 0 to 210 seconds.
   (b) What is its average velocity in that time interval?
   (c) What is the average acceleration in the same time interval?

17

Arc Length and *Mathematica* Procedures

> A projectile is fired into the air at an angle to the horizontal. What is the length of the path it follows?

In order to answer questions like these, we need to be able to find the length of a curved path. For a specific example of projectile motion to look at, suppose that a projectile's position is given by the parametric equations:

$$x = 100\,t$$
$$y = -16\,t^2 + 320\,t$$

where x and y are in feet, and t is in seconds. To get an idea of the motion, we'll plot the curve, but first we need to know how long it takes the projectile to return to earth.

```
In[1]:=
      Clear[x,y,t];
      x[t_]:= 100t;
      y[t_]:= -16t^2 + 320t;
      Solve[y[t]==0, t]

Out[1]=
          {{t -> 20}, {t -> 0}}
```

The projectile starts on the ground at $t = 0$ seconds and is back on the ground at $t = 20$ seconds. Now set up the plot command to show the entire path of motion.

```
In[2] :=
        ParametricPlot[{x[t],y[t]},   {t,0,20}];
```

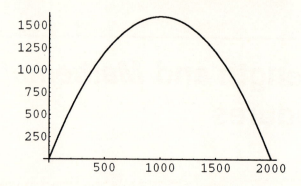

The question is this: Exactly how many feet are in the arched path that the projectile follows? We can approximate the distance travelled by finding points on the curve and pretending that the projectile travels in a straight line between the points. Suppose that we take three points on the path of the projectile, the points with $t = 0$, 10, and 20, and assign the name **points** to the list containing them.

```
In[3] :=
        points  =  Table[{x[i],y[i]},{i,0,20,10}]
```

```
Out[3]=
        {{0,  0},  {1000,  1600},  {2000,  0}}
```

We can see the path together with the points and the lines joining them using this form of the **ParametricPlot** command.

```
In[4] :=
        ParametricPlot[
          {x[t],y[t]},  {t,0,20},
          Epilog->{{
            PointSize[.03],
            Map[Point,points],
            Line[points]
            }}
          ];
```

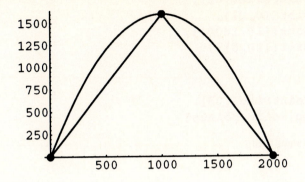

About Map and Lists

In just a minute we will calculate the length of those two line segments, but first we need to talk a little about that setting, in particular, about the effect of **Map**. Previously, we would have written this **Epilog** setting as

```
Epilog->{{
   PointSize[.02],
   Point[{0,0}],
   Point[{10,1600}],
   Point[{20,0}],
    Line[{{0,0},{1000,1600},{2000,0}}]
   }}
```

Notice that the argument of the **Line** command above is the list **points**. So, why retype the list? **Line[points]** is the same command. The elements of the list **points** also appear as the arguments of the three **Point** commands. **Map[Point,points]** applies the function **Point** to each element in **points**. Enter this command by itself to see the effect of **Map**.
   In[5]:=

```
Map[Point,points]
```

Out[5]=
```
       {Point[{0,  0}], Point[{1000,  1600}],
          Point[{2000,  0}]}
```

Map is necessary because the command **Point** accepts only individual points as arguments, not lists of points. **Line** can be applied to a list, but **Point** cannot. In any case,

```
{PointSize[.02],
Point[{0,0}],
Point[{10,1600}],
Point[{20,0}]}
```

and

```
{PointSize[.03],
Map[Point,points]}
```

say exactly the same thing.

A Distance Function

The distance between two points (a, b) and (c, d) can be calculated using the familiar Pythagorean distance formula $((c-a)^2 + (d-b)^2)^{1/2}$. Since we will be calculating a lot of distances in the exploration, it will be convenient to have a function which makes the distance calculation for us.

```
In[6]:=
        distance[{a_,b_},{c_,d_}]:=
          N[Sqrt[(c - a)^2 + (d - b)^2]]
```

Use this distance function to find the total lengths of the two lines drawn earlier.

```
In[7]:=
        d1 = distance[{0,0},{1000,1600}]
Out[7]=
        1886.8
```

```
In[8]:=
        d2 = distance[{1000,1600},{2000,0}]
Out[8]=
        1886.8
```

The total length will be the sum of these two distances, 3773.59 feet.

```
In[9]:=
        d1 + d2
Out[9]=
        3773.59
```

From the graph it is fairly clear that the actual length of the curve is greater than the value we just found. By using more points we can get a closer approximation. The next set of commands finds the positions of the projectile at $t = 0, 5, 10, 15,$ and 20 seconds and draws the line segments between them.

```
In[10]:=
        Clear[points];
        points = Table[{x[t],y[t]},  {t,0,20,5}];
        ParametricPlot[
          {x[t],y[t]},{t,0,20},
          Epilog->{{
            PointSize[.03],
            Map[Point,points],
            Line[points]
            }}
          ];
```

Compute the four distances and total them.

```
In[11]:=
        d1 = distance[{x[0],y[0]},{x[5],y[5]}]

Out[11]=
        1300.

In[12]:=
        d2 = distance[{x[5],y[5]},{x[10],y[10]}]

Out[12]=
        640.312

In[13]:=
        d3 = distance[{x[10],y[10]},{x[15],y[15]}]

Out[13]=
        640.312
```

```
In[14]:=
        d4 = distance[{x[15],y[15]}, {x[20],y[20]}]
```

```
Out[14]=
        1300.
```

```
In[15]:=
        length = d1 + d2 + d3 + d4
```

```
Out[15]=
        3880.62
```

This estimate, 3880.62 feet, is closer to the actual length of the projectile's path than our previous one was, and the estimate can be improved still further by taking more points. And we will do that. However, this is a situation which begs for automation, since the sequence of operations and the nature of the calculations made at each step are identical. The only thing that changes as we improve the estimate is the number of points. Otherwise, the outline of the steps goes like this:

1. Pick a set of time values between 0 seconds and 20 seconds.
2. Find the position of the projectile at each of the times chosen.
3. Calculate the straight line distance from each position to the next.
4. Add the distances together to get the approximate length of the path.

The idea is to write a set of *Mathematica* commands which will perform all of these steps automatically. We want *Mathematica* to do all the work once we tell it how many time values to use. Let's begin by thinking of the number of time values after the initial time as n, and, for convenience, let's space them out equally between the beginning time 0 and the end time 20, so the interval between one time and the next will be $dt = 20/n$.

We will eventually tell *Mathematica* about the decisions made so far by entering the following code, but before doing so let's take a close look at the function **t**.

```
        Clear[t];
        dt = 20/n;
        t[0] = 0;
        t[m_]:= t[m - 1] + dt
```

The equation **t[0] = 0** tells *Mathematica* that the initial time value is zero. Subsequent time values are to be calculated by the definition **t[m_]:= t[m - 1] + dt**. This definition says that the time for the next position of the projectile is calculated by adding the time interval dt to the preceding time value. So, t_1 would be found as follows: $t_1 = t_0 + dt$ and t_2 by the calculation $t_2 = t_1 + dt$ and so forth. Functions which make their calculations in this fashion by referring back to previous function values are called *recursive* functions. The *Mathematica* function **t[m_]:= t[m - 1] + dt** is recursive.

We could have gotten away with a more straightforward definition of time, something like **t[m_]:= dt*m**, which is not recursive; but recursive functions are important, so we might as well take the opportunity to get used to them here.

As an example, let's see what these instructions produce if $n = 4$. In this case, dt will be 5, and so the time values should be

$$t_0 = 0$$
$$t_1 = t_0 + dt = 0 + 5 = 5$$
$$t_2 = t_1 + dt = 5 + 5 = 10$$
$$t_3 = t_2 + dt = 10 + 5 = 15$$
$$t_4 = t_3 + dt = 15 + 5 = 20$$

These are exactly the same time values that we used earlier. Let's see if the definition returns them by entering the definition of t and setting n to 4.

```
In[16]:=
        Clear[t];
        n = 4;
        dt:= 20/n;
        t[0] = 0;
        t[m_]:= t[m - 1] + dt;
        Table[t[k], {k,0,4}]

Out[16]=
        {0, 5, 10, 15, 20}
```

Perfect! Now that we have a function which will generate the time values between 0 and 20, we need another command to generate the points on the path corresponding to these times. Enter the command below. Note that *Mathematica* is still thinking of n as 4, so the output should be a set of 5 points on the path of the projectile.

```
In[17]:=
        points = Table[{x[t[k]], y[t[k]]},
                       {k,0,n}]

Out[17]=
        {{0, 0}, {500, 1200}, {1000, 1600},
         {1500, 1200}, {2000, 0}}
```

Good. Steps one and two are completed. Move on to three. It calls for a command which will calculate the distance from a position listed in a table like the one above to the next position. How about this?

```
In[18]:=
        d[k_]:= distance[
            {x[t[k-1]], y[t[k-1]]},
            {x[t[k]], y[t[k]]}
            ]
```

Give it a try. The distances it returns for $k = 1, 2, 3, 4$ should be the same four distance values we calculated earlier.

```
In[19]:=
        Table[d[k], {k,1,4}]
```

```
Out[19]=
        {1300., 640.312, 640.312, 1300.}
```

We are definitely on a roll! Looks like the **d** function works.

Now, for the total distance, just sum the 4 values of **d**.

```
In[20]:=
        Sum[d[k], {k,1,n}]
```

```
Out[20]=
        3880.62
```

Exactly the value we found earlier.

A Procedure for Approximating the Arc Length of the Projectile's Path

There is one final step in the process of automation: We must gather all of the *Mathematica* code in a single cell where it can executed all at once. This is done below. We have also tossed in a **Parametric Plot** command so the procedure will draw a picture of the path and the approximating line segments. A **Clear** command is inserted at the beginning so variable assignments don't get all mixed up. You will also notice that the three sections of code are introduced by descriptive comments. These headings are set off by asterisks and enclosed in parentheses so that *Mathematica* will understand that they are only comments and are not instructions to be sent to the kernel.

```
In[21]:=
        (*define and initialize variables*)

        Clear[x,y,dt,t,points,distance,d,length];
        x[t_]:= 100 t;
        y[t_]:= - 16 t^2 + 320 t;
        n = 5;
        dt = 20/n;
        t[0] = 0;
        t[m_]:= t[m] = t[m - 1] + dt;
        points = Table[
                {x[t[k]], y[t[k]]}, {k,0,n}
                ];
```

```
(*plot the path of the projectile*)

ParametricPlot[
    {x[z],y[z]}, {z,0,20},
    PlotStyle->{{
       PointSize[.03],
       Map[Point,points],
       Line[points]
       }}
    ];

(*calculate the length of the path*)

distance[{a_,b_},{c_,d_}]:=
    N[Sqrt[(c - a)^2 + (d - b)^2]];

d[k_]:= distance[
        {x[t[k-1]],y[t[k-1]]},
        {x[t[k]],y[t[k]]}
        ];
Sum[d[k],  {k,1,n}]
```

Select the entire cell containing the procedure and enter it.

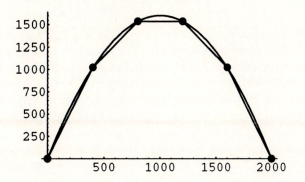

```
Out[21]=
      3898.16
```

Beautiful! It passes the test.

There is one new thing in the procedure code. Note the line that defines **t[m]**. It is

$$t[m\_]:= t[m] = t[m - 1] + dt$$

This form of a recursive definition can speed up the calculation of function values significantly. It causes *Mathematica* to automatically store the value of **t[m]** at the time it is calculated. If you give the command in the form described previously,

$$t[m\_]:= t[m - 1] + dt$$

then when you ask for **t[5]**, for example, *Mathematica* will have to compute **t[4]**, **t[3]**, **t[2]**, and **t[1]** again, even if it has already found all those values. The new form of the command causes each **t** values to be saved as soon as it is calculated, so it will not have to be recomputed.

Practice Exercise 1: Copy, **Paste**, and reenter the procedure with **n** set to a different value.

Practice Exercise 2: An elliptical race track has a major axis of length 100 meters and a minor axis of length 50 meters and so can be described by the parametric equations $x(t)$ = 50 cos(t) and $y(t)$ = 25 sin(t), $0 \leq t \leq 2\pi$. Edit the Arc Length procedure to estimate the length of one lap. (*Ans.* about 242 feet)

Finding Arc Length Using Integration

If you look up arc length in a calculus book, you will find that there is an integral formula for computing it. It probably looks something like this:

$$\int (x'(t)^2 + y'(t)^2)^{1/2} dt$$

where the interval of integration is the time interval. So, why didn't we just give you the formula to begin with? There are two reasons: (1) The work we have done here will help you to understand what arc length is and where the formula comes from. (2) Many (in fact, most) of the integrals which arise in arc length problems can't be evaluated exactly since their integrands do not have elementary antiderivatives. Consequently, they require an approximation technique for solution anyway. So, why not think about numerical arc length calculations as approximations from the beginning?

　　The projectile problem that began this chapter is an example of an arc length calculation which *can* be solved by antidifferentiation. Here are the derivatives of $x(t)$ and $y(t)$:

```
In[22]:=
        Clear[x,y,t]
        x[t_]:= 100t
        y[t_]:= -16t^2 + 320t

In[23]:=
        x'[t]
```

```
Out[23]=}
        100
```

```
In[24]:=
        y'[t]
```

```
Out[24]=
        320 - 32 t
```

Ask *Mathematica* for the antiderivative of $(x'(t)^2 + y'(t)^2)^{1/2}$.

```
In[25]:=
        Integrate[Sqrt[100^2 + (320-32t)^2],t]
```

```
General::load: Loading package ComplexExpand'.
```

```
Out[25]=
              t
        (-5 + -) Sqrt[112400 - 20480 t + 1024 t² ] +
              2

                      8 (-10 + t)
          625 ArcSinh[───────────]
          ─────────────────25───────
                      4
```

Now, evaluate the antiderivative at $t = 20$ and again at $t = 0$ and take the difference.

```
In[26]:=
        N[(%/.t->20) - (%/.t->0)]
```

```
Out[26]=
        3940.07
```

Rounded to the nearest foot this is 3940 feet, which is exactly the result we got earlier using line segment approximations.

Now, try the integral formula on the length of that elliptical track described in the practice exercise.

```
In[27]:=
        Clear[x,y,t]
        x[t_]:= 50Cos[t]
        y[t_]:= 25Sin[t]
```

```
In[28]:=
        x'[t]
```

```
Out[28]=
        -50 Sin[t]
```

```
In[29]:=
        y'[t]
```

```
Out[29]=
        25 Cos[t]
```

Here is the arc length integral:

```
In[30]:=
        Integrate[
          Sqrt[(-50Sin[t])^2 + (25Cos[t])^2],
          t]
```

```
Out[30]=
                          3125 - 1875 Cos[2 t]
        Integrate[Sqrt[————————————————————————], t]
                                  2
```

Mathematica is telling you that it can't find an exact antiderivative. The truth is that none exists. **NIntegrate** will give a result, but it returns an approximation to the integral much like the approximations you have already calculated in the practice exercise.

```
In[31]:=
        NIntegrate[
          Sqrt[(-50Sin[t])^2 + (25Cos[t])^2],
          {t,0,2Pi}
          ]
```

```
Out[31]=
        242.211
```

An Optional Section for Those Interested in a Further Automation of the Arc Length Procedure

The arc length procedure we developed above is a fine one, but it can be improved. It is limited in at least one important way: The time interval is fixed at 0 to 20 seconds. What if you wanted the length of only a portion of the path, say from 10 seconds to 15 seconds? It would be nice to have a superflexible procedure which could adapt to changes in the time interval. Of course such a thing is possible. Here it is:

```
In[32]:=
        ArcLength[x_,y_,e_,f_,n_]:=
        Block[
          {dt,t,points,distance,d,m,k},

        (*define and initialize variables*)

          dt = N[(f-e)/n];
          t[0]  = e;
          t[m_]:= t[m] = t[m - 1] + dt;

          points = Table[
            {x[t[k]], y[t[k]]}, {k,0,n}
            ];

        (*plot the path of the projectile*)

          ParametricPlot[
            {x[z],y[z]}, {z,e,f},
            PlotStyle->{{
              PointSize[.03],
              Map[Point,points],
              Line[points]
              }}
            ];

        (*calculate the length of the path*)

          distance[{a_,b_},{c_,d_}]:=
            N[Sqrt[(c - a)^2 + (d - b)^2]];
          d[k_]:= distance[
            {x[t[k-1]], y[t[k-1]]},
            {x[t[k]], y[t[k]]}
            ];
          N[Sum[d[k],{k,1,n}]]
          ]
```

You see that the code differs only very slightly from the procedure we developed in this exploration. The effect of the changes in code, however, is enormous. The entire procedure is now a function, **ArcLength**, which has five arguments, the two parametric equations for the path of motion, the beginning time, the ending time, and the number of approximating line segments. Now, if you wish to find arc length you do not need to edit the procedure for each application. Just enter the function definition once and then use the function for each of the problems.

We should test this function. Apply it to the elliptical path from Practice Exercise 2, using an approximation based on 8 segments.

```
In[33]:=
        Clear[x,y,t]
        x[t_]:= 50Cos[t]
        y[t_]:= 25Sin[t]
        ArcLength[x,y,0,2Pi,8]
```

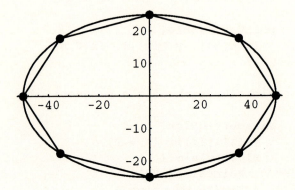

```
Out[33]=
        236.245
```

It works!

A word about the code in the definition of **ArcLength**. The **Block** command is a new one. Its effect is to make the variables in the list following it into what are called *local* variables. This means that any values previously assigned to these variables are automatically cleared before the procedure is executed, and further, that the earlier assignments are restored when the procedure is completed. Another change: The definition of **dt** uses **e** and **f** for the beginning and ending times instead of 0 and 20, and **t[0]** is set to **e.** In the **ParametricPlot** command, the plot parameters are changed from {**z, 0, 20**} to {**z, e, f**}. These small adjustments are all that is needed to allow for a variable time interval.

If you wish to use this function, you can copy the code from the Input file on your *Exploring Calculus* disk.

Code for the Practice Exercises

2. For example,

```
(*define and initialize variables*)

   Clear[x,y,dt,t,points,distance,d,length];
   x[t_]:= 50Cos[t];
   y[t_]:= 25Sin[t];
   n = 8;
   dt = N[2Pi/n];
   t[0] = 0;
   t[m_]:= t[m] = t[m - 1] + dt;


   (*plot the path of the projectile*)

   points = Table[
     {x[t[k]], y[t[k]]}, {k,0,n}
     ];
   ParametricPlot[{x[z],y[z]},  {z,0,2Pi},
     PlotStyle->{{
       PointSize[.03],
       Map[Point,points],
       Line[points]
       }}
     ];


   (*calculate the length of the path*)

   distance[{a_,b_},{c_,d_}]:=
     N[Sqrt[(c - a)^2 + (d - b)^2]];
   d[k_]:= distance[
     {x[t[k-1]], y[t[k-1]]},
     {x[t[k]],y[t[k]]}
     ];
   Sum[d[k],  {k,1,n}]
```

Problems

1. Find the length of one arch of the cycloid

$$x = 5(t - \sin(t)),$$
$$y = 5(1 - \cos(t)).$$

   (a) using approximations based on line segments;
   (b) using integration. (**NIntegrate** works better than **Integrate** in these problems.)

2. Find the length of the Archimedean spiral for $0 \le t \le 2\pi$

$$x = t \cos(t),$$
$$y = t \sin(t)$$

   (a) using approximations based on line segments;
   (b) using integration.

3. Find the length of the curve $y = x^3$ from $x = 0$ to $x = 1$. Describe the curve parametrically with the equations $x = t$ and $y = t^3$ and use both integration and approximations based on line segments. (Use the **NIntegrate** command.)

4. What number does the length of the curve $y = x^n$ for $0 \le x \le 1$ approach as n becomes infinitely large? Print a sequence of plots illustrating your answer, and write a paragraph of explanation.

18

Euler's Method

When you take a derivative, you get information about the way a curve is changing. For example, if the derivative at a point is 2, then in a small neighborhood of the point the curve behaves like a line segment with slope 2, so you know that as the curve passes through the point it is rising less rapidly than it would if the slope were, for example, 10.

You have had experience using a derivative to figure out how a function behaves. In this exploration we are going to look at the reverse situation and consider problems in which the derivative of a function is known but the function itself is not. Just in case you imagine such a situation to be unlikely, suffice it to say that most applications of the calculus involve differential equations, formulas containing information about a function's derivative. Such an equation is solved when the function is found.

Sometimes it is a relatively simple matter to move "backwards" from the derivative to the function from which it was derived. For example, suppose you know that $f'(x) = x^2$ and further that $f(0) = 0$. Then it is an easy matter to determine another function value, say $f(3)$. The derivative tells you that $f(x) = (1/3)x^3 + C$ and C must be 0 in order for $f(0)$ to equal 0. Therefore, the function is $f(x) = (1/3)x^3$ and consequently $f(3) = (1/3)3^3 = 9$.

In this example, you were able to figure out $f(x)$ from derivative facts you know, but often that is not possible or, at best, difficult. Consider, for example, a function $f(x)$ whose derivative is $((16-3x^2)/(16-4x^2))^{1/2}$. Suppose you wish to know $f(3)$. In this case, its much more difficult to go "backwards" from the derivative $f'(x) = ((16-3x^2)/(16-4x^2))^{1/2}$ to the formula for $f(x)$. Even *Mathematica* cannot handle the problem directly.

```
In[1]:=
        Integrate[Sqrt[(16 - 3x^2)/(16 - 4x^2)],x]
```

```
Out[1]=
                             16 - 3 x²
             Integrate[Sqrt[————————————], x]
                             16 - 4 x²
```

No luck; and without an exact antiderivative the problem of computing a specific function value, like $f(3)$, is challenging. With *Mathematica*'s help we can take up the challenge. Think about it in the context of a specific problem.

Given that $g'(x) = x^2 + 1$ and that $g(0) = 3$, find g(2).

Admittedly, this problem is not so hard. You can probably figure out the rule for $g(x)$. But, for the time being, pretend that you can't, and let's see how you might construct an approximation to $g(2)$ using the information you have about g's derivative. The technique we will use is called Euler's Method. When you understand how it works on this simple problem, we will apply it to functions whose antiderivatives aren't known. Meanwhile, because we can figure out the exact solution, we will have a way to check our work as we go along.

Begin by teaching *Mathematica* the formula for the derivative. Call it **gp** for g-prime.

```
In[2]:=
        gp[x_]:= x^2 + 1
```

We know that the value of $g(x)$ is 3 when x is zero, and we are being asked to determine $g(2)$, so we need to find out as much as possible about the behavior of this function on the interval from $x = 0$ to $x = 2$. One thing we know for sure is that the curve is rising, since the derivative is positive, $((x^2 + 1) > 0$ for all $x)$. We can compute its second derivative.

```
In[3]:=
        D[gp[x],x]
```

```
Out[3]=
        2 x
```

This expression will also be positive for $0 \le x \le 2$, and so the curve is concave up over this interval.

Piecing this information together, we know that the curve is increasing and concave up for $0 \le x \le 2$, so the graph will look something like this:

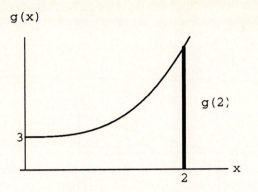

The height of the curve over the number 2 on the *x*-axis is $g(2)$. We're going to use Euler's Method to estimate this height. It will involve drawing a much more precise graph of $g(x)$, one that takes into account the fact that we know the value of $g'(x)$ for each *x* between 0 and 2.

As we mentioned earlier, we need a way to test our results, so pause for a minute and compute the exact value for $g(2)$ using antidifferentiation.

In[4] :=

```
        Integrate[gp[x],x]
```

Out[4]=

$$x + \frac{x^3}{3}$$

In order to get the particular *g* function we are studying here, for which $g(0) = 3$, we must add the constant 3 to the expression for the antiderivative. So, $g(x) = x + x^3/3 + 3$, and $g(2) = 23/3 = 7.66667$.

Keep that number under your hat because we are pretending, for the purposes of studying Euler's Method, that we aren't able to calculate it.

A Rough Approximation to g(2)

Here is a very crude way to approximate $g(2)$: Assume that the curve $y = g(x)$ has the same slope throughout the interval $0 \le x \le 2$ that it has when $x = 0$. The slope of $g(x)$ at 0 is given by **gp**.

In[5] :=

```
        gp[0]
```

Out[5]=

```
        1
```

This output tells us that the line tangent to the curve $g(x)$ at the point $(0, 3)$ has a slope of 1. Here is a plot of this tangent line over the entire interval from 0 to 2:

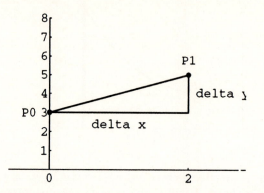

The distance from 2 on the x-axis to the point P_1 is our first approximation to $g(2)$. This distance is equal to the second coordinate of the P_1 and can be found by adding Δy to 3. We just have to figure out the value of Δy. No problem, since $\Delta y/\Delta x = g'(0) = 1$. Here are the details of the calculation:

$$g(2) \approx 3 + \Delta y = 3 + 1*\Delta x = 3 + 1*2 = 5$$

Before we go on to refine this admittedly very rough stab at a value for $g(2)$, we need some terminology. Look at the plot. The point P_0 is called the *initial point*. Its coordinates are symbolized $(x(0), y(0))$. In this plot, $x(0) = 0$ and $y(0) = 3$. The point P_1 is the first point generated in this Euler approximation. Its coordinates are symbolized $(x(1), y(1))$, so $x(1) = 2$ and $x(1) = 5$. In this case, the first point was the last point, and so its second coordinate provides the approximation to $g(2)$.

$$g(2) \approx y(1) = 5$$

A Better Approximation

The first approximation was a crude one because we assumed that the slope of $g(x)$ remained constant as x increased from 0 to 2. This is not correct, because the slope increases as x increases. The tangent to the curve at 0 will approximate the curve well for only a short distance, after which the curve bends away from the line. Euler's Method takes this into account and applies the slope at zero to only the first part of the interval between 0 and 2. Then it uses the formula $g'(x)$ to make a slope correction which it applies for a second short interval, after which another correction is made. This process continues until the small intervals cover the distance from 0 to 2. If four slopes were used, the one at 0 and three corrections, the picture would look something like this:

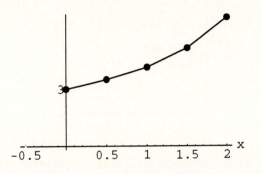

The vertical distance from (2, 0) to the graph is the approximation to $g(2)$. Let's get *Mathematica* to make this calculation for us. It won't be easy even with *Mathematica* to help, but once you see how Euler's Method works in this case it will be a simple matter to automate Euler Approximations, which will make all subsequent calculations a breeze.

In the picture above, a slope correction was made after every one-half unit increase in x. This means that beginning at P_0 four new points were generated. We are going to name them, P_1, P_2, P_3, P_4. The first coordinate of P_0 is 0. So we have, as before, $x(0) = 0$. The first coordinates of the four new points are a half unit apart. Consequently, we have $x(1) = 0.5, x(2) = 1.0, x(3) = 1.5$, and $x(4) = 2.0$. Teach *Mathematica* the rule for these first coordinates.

```
In[6]:=
        Clear[x,k,dx]
        dx = 0.5
        x[0] = 0
        x[k_]:= x[k] = x[k-1] + dx
```

Make sure that *Mathematica* has it right.

```
In[7]:=
        Table[x[k], {k,0,4}]

Out[7]=
        {0, 0.5, 1., 1.5, 2.}
```

Good. Now, let's think about how we will calculate the y-coordinates of the four points we want to generate. We know that $y(0) = 3$. We'll figure out $y(1)$ using Euler's Method as we did earlier: Assume that the rate at which the function is increasing when $x = 0$ remains constant over the interval from $x = 0$ to $x = 0.5$. The picture looks like this:

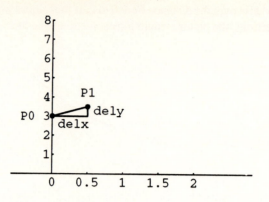

Calculate the second coordinate of P_1. Recall that the slope at P_0 is 1.

$$y(1) = 3 + \Delta y = 3 + \Delta x * 1 = 3 + 0.5 = 3.5.$$

Now we know that P_1 has coordinates $x(1) = 0.5, y(1) = 3.5$. The next step is to find the second point, P_2. We already know that its first coordinate is $x(2) = 1.0$. Its second coordinate takes a little work. First, we let the formula g' tell us what the slope of the curve is when $x = 0.5$.

In[8]:=
          gp[.5]

Out[8]=
          1.25

Then, we mark off a line segment from P_1 having that slope and a run of $\Delta x = 0.5$. Last, we will add the rise of this segment to $y(1)$ to get $y(2)$. Here is the picture:

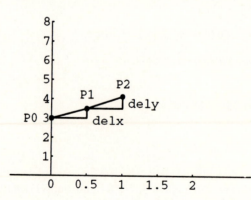

The slope of the segment from P_1 to P_2 is 1.25, so

$$\Delta y/\Delta x = 1.25$$
$$\Delta y = \Delta x * 1.25 = 0.5 * 1.25 = 0.625$$
$$y(2) = y(1) + 0.625 = 3.5 + 0.625 = 4.125$$

This is progress. We know that $x(2) = 1.0$, $y(2) = 4.125$.

Practice Exercise 1: Below is the picture for P_3. Make a slope correction at P_2 and assume that the new rate of change, $g'(1.0)$, applies for x between 1.0 and 1.5. Calculate $y(3)$. (*Ans.* 5.125)

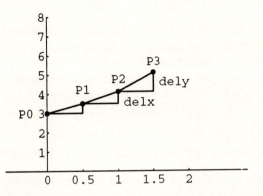

Practice Exercise 2: Below is the picture for P_4. Calculate $y(4)$. (*Ans.* 6.75)

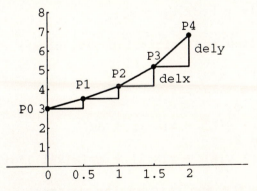

This final picture shows the approximation to $g(2)$. It is the distance of the point P_4 above 2 which is $y(4) = 6.75$. Notice that this number is much closer to the true value of $g(2)$, 7.67, than our first approximation was. The approximation improved because we used four slope values instead of just one.

We could do even better by making more slope corrections. The key is to use a much smaller Δx and generate many more points so that the slopes of the segments from one point to the next more closely fit the curve of g's graph. For example, if Δx were 0.1,

Euler's Method would generate 20 points based on 20 different slope values. With $\Delta x = 0.01$ the method would use 200 slope values to generate 200 points. These are not calculations that anyone wants to do by hand or even with a calculator. It's a job for a *Mathematica* procedure.

Writing the procedure is simply a matter of teaching *Mathematica* how to calculate the x and the y coordinates of the P points.

We already have a procedure for calculating the x-coordinates.

```
Clear[x,k,dx];
dx = 0.5;
x[0] = 0;
x[k_]:= x[k] = x[k-1] + dx
```

The only part of it we will have to change is the number assigned to **dx**. The more slope changes we make between 0 and 2 the smaller **dx** must be.

The procedure for finding the y-coordinates mimics the way in which y-coordinates were found in the example; the y-coordinate of a point was found by adding Δy to the preceding point's y-coordinate. In turn, Δy was calculated by multiplying the slope at the preceding point by Δx. This bit of *Mathematica* code says it all.

```
y[0] = 3;
y[k_]:= y[k] = y[k-1] + gp[x[k-1]]*dx
```

Now we are ready to put together a procedure which will solve the example problem. If it works, we can figure out how to edit it in order to generate a better approximation. The following code will generate the points and display them in table form:

```
In[9]:=
    Clear[x,y,k,dx,gp,pts];

    (* define and initialize variables*)

    gp[x_]:= x^2 + 1;
    dx = .5;
    x[0] = 0;
    x[k_]:= x[k] = x[k-1] + dx;
    y[0] = 3;
    y[k_]:= y[k] = y[k-1] + gp[x[k-1]]*dx;

    (*generate and display the points*)

    pts = Table[{x[k],y[k]}, {k,0,4}];
    TableForm[
      pts,
      TableHeadings->{
        {0,1,2,3,4}, {"x\n","y\n"}
        }
      ]
```

```
Out[9]//TableForm=
          x       y

     0    0       3
     1    0.5     3.5
     2    1.      4.125
     3    1.5     5.125
     4    2.      6.75
```

Perfect! Look at a plot of **pts**.

```
ListPlot[
  pts,
   PlotStyle->{PointSize[.02]},
   PlotRange->{0,8}
   ];
```

How should the procedure be changed in order to generate more points? Well, the answer depends in part on how many points you want: 20? 100? Both? Why not keep our options open by editing the procedure for a general number of points *n* like this:

```
In[10]:=
        n = 20
        Clear[x,y,k,dx,gp,pts];

        (* define and initialize variables*)

        gp[x_]:= x^2 + 1;
        dx = N[2/n];
        x[0] = 0;
        x[k_]:= x[k] = x[k-1] + dx;
        y[0] = 3;
        y[k_]:= y[k] = y[k-1] + gp[x[k-1]]*dx;
```

```
(*generate and display the points*)

pts = Table[{x[k],y[k]}, {k,0,n}];
ListPlot[
  pts,
  PlotStyle->{PointSize[.02]},
  PlotRange->{0,8}
  ];
"g("<>ToString[x[n]]<>") ≈ "
  <>ToString[N[y[n]]]
```

```
Out[10]=
        g(2) ≈ 7.47
```

If you want to use more than 20 intervals, just change the value assigned to n in the first line.

The last two lines of code are a new addition. They cause the approximation to $g(2)$ to be displayed. The approximation printed out is fairly close to the true value. Remember that the exact value is 7.67. The approximation is better than our previous one because these points fit the g curve better than the four-point approximation did. Here is the real g curve superimposed on the plot above:

```
In[11]:=
        Plot[
            x^3/3 + x + 3, {x,0,2},
            Epilog->{PointSize[.02], Map[Point,pts]}
            ];
```

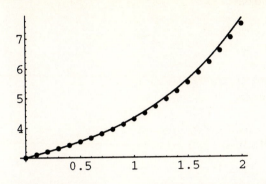

The fit is pretty good, although the curve does bend away from the points a bit. And this should not be surprising. The slope of the curve changes continuously, while the points generated by the Euler approximation were found using only 20 of these slopes.

Practice Exercise 3: Approximate $g(3)$ using $y(30)$. How close is this number to the real value of $g(3)$? Plot the 30 points superimposed on the plot of g. (*Ans.* 14.555)

Code for the Practice Exercise

3. Use **n = 30** and **dx= 3/n**. Otherwise the procedure remains the same.

Problems

1. Suppose $g'(x) = x^2$ and $g(0) = 0$.
   (a) Calculate $g(x)$ using antidifferentiation. Use $g(x)$ to find $g(1)$.
   (b) Using $\Delta x = 0.1$ and the starting point $x(0) = 0$, $y(0) = 0$ calculate a table of points, **pts**, which approximates the graph of $y = g(x)$. Plot **pts** and $g(x)$.
   (c) How close is $y(10)$ to the correct value of $g(1)$?
   (d) Use $\Delta x = 0.01$ and find an approximate value for $g(1)$. How close is it to the true value?

2. Suppose that $g'(x) = 1/x$ and $g(1) = 0.5$. Find $g(2)$.
   (a) The starting point is $x(0) = 1$, $y(0) = 0.5$. Use $\Delta x = 0.1$, and set up the commands to find $x(k)$ and $y(k)$.
   (b) Make a table of the points and plot the points.

3. Let $f'(x) = ((1 + 2x^2)/(1 - 2x^2))^{1/2}$ and $f(0) = 0$. Use Euler's Method for $n = 2, 4, 8, 16,...$, until you find two which agree to one decimal place on the approximate value of $f(0.5)$.

4. Suppose that $g'(x) = (1 + x^2)^{1/2}$ and $g(0) = 0$. Use Euler's Method for $n = 2, 4, 8, 16,...$, until you find two which agree to one decimal places on the approximate value of $f(1.5)$.

19

The Fundamental Theorem of Calculus, *An Informal View*

In this chapter we'll take a second look at Euler's Method. There is more there than meets the eye.

Finding *y*(20) Using Sum

Recall the procedure we devised for finding $g(2)$ given that $g'(x) = x^2 + 1$ and $g(0) = 3$.

```
In[1]:=
    n = 20;

    Clear[x,y,k,dx,gp,pts];

    (* define and initialize variables*)

    gp[x_]:= x^2 + 1;
    dx = N[2/n];
    x[0] = 0;
    x[k_]:= x[k] = x[k-1] + dx;
    y[0] = 3;
    y[k_]:= y[k] = y[k-1] + gp[x[k-1]]*dx;
```

```
(*generate and display the points*)

pts = Table[{x[k],y[k]}, {k,0,20}];
ListPlot[
    pts,
    PlotStyle->{PointSize[.02]},
    PlotRange->{0,8},
    AxesLabel->{"x","g(x)"},
    PlotLabel->"y = g(x)"
    ];
"g(2) ≈ y[20] = "<>ToString[y[20]]
```

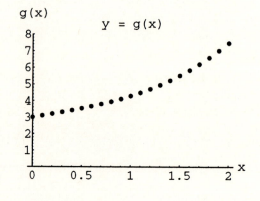

Out[1]=

$$g(2) \approx y[20] = 7.47$$

Here is the new wrinkle: Sometimes in calculating Euler approximations for function values, you will not need to generate the entire list of points. You will only be interested in the "bottom line," the second coordinate of the last generated point. In our example above this would be the number $y(20)$. When this is the case, there is a quick way to get to the number you want.

To see how it works, you need to know exactly how the slope formula **gp** was used to compute $y(20)$. There is a very clever way to get this information. First, **Copy** the first seven lines of the procedure above and **Paste** them into a new cell in your notebook. When you have done that, delete the line which defines **gp**. Now, select the procedure and *enter* it. The **Clear** command will remove **gp** from *Mathematica*'s memory. The rest of the definitions will remain as before. Finally, ask for **y[20]**.

```
In[2]:=
        Clear[x,y,k,dx,gp];
        dx = N[2/n];
        x[0] = 0;
        x[k_]:= x[k] = x[k-1] + dx;
        y[0] = 3;
        y[k_]:= y[k] = y[k-1] + gp[x[k-1]]*dx;
        y[20]

Out[2]=
        3 + 0.1 gp[0] + 0.1 gp[0.1] + 0.1 gp[0.2] +
          0.1 gp[0.3] + 0.1 gp[0.4] + 0.1 gp[0.5] +
          0.1 gp[0.6] + 0.1 gp[0.7] + 0.1 gp[0.8] +
          0.1 gp[0.9] + 0.1 gp[1.] + 0.1 gp[1.1] +
          0.1 gp[1.2] + 0.1 gp[1.3] + 0.1 gp[1.4] +
          0.1 gp[1.5] + 0.1 gp[1.6] + 0.1 gp[1.7] +
          0.1 gp[1.8] + 0.1 gp[1.9]
```

What is returned is not a number, but a formula. *Mathematica* can't calculate a number because it no longer knows a definition for **gp**, so it does the next best thing and tells you how it would have used **gp** had it known it. Next best for us is just perfect, because we can use the information in the formula to devise a command which will compute y[20] directly. The idea is to construct a **Sum** command which produces the identical formula. Not difficult at all. Here it is.

```
In[3]:=
        3 + Sum[gp[k*0.1]*0.1, {k,0,19}]

Out[3]=
        3 + 0.1 gp[0] + 0.1 gp[0.1] + 0.1 gp[0.2] +
          0.1 gp[0.3] + 0.1 gp[0.4] + 0.1 gp[0.5] +
          0.1 gp[0.6] + 0.1 gp[0.7] + 0.1 gp[0.8] +
          0.1 gp[0.9] + 0.1 gp[1.] + 0.1 gp[1.1] +
          0.1 gp[1.2] + 0.1 gp[1.3] + 0.1 gp[1.4] +
          0.1 gp[1.5] + 0.1 gp[1.6] + 0.1 gp[1.7] +
          0.1 gp[1.8] + 0.1 gp[1.9]
```

This is the plan: We will reteach *Mathematica* the formula for **gp** and follow it with the **Sum**. The result should be *y*(20). Let's see.

```
In[4]:=
        gp[x_]:= x^2 + 1;
        3 + Sum[gp[k*0.1]*0.1, {k,0,19}]

Out[4]=
        7.47
```

We obviously have a winner.

Try another example: Find $h(4)$ given that $h(1) = 0$ and $h'(x) = 1/x$. Using a **Sum** command based on, say, 20 slopes. The changes from the above example are: $dx = (4-1)/n = 3/n$; **gp** is replaced by **hp** everywhere; x[0] = 1; and y[0] = 0. First enter the procedure without a definition for **hp**.

```
In[5]:=
        n = 20;
        Clear[x,y,k,dx,hp];
        dx = N[3/n];
        x[0] = 1;
        x[k_]:= x[k] = x[k-1] + dx;
        y[0] = 0;
        y[k_]:= y[k] = y[k-1] + hp[x[k-1]]*dx;
```

Ask for y[20].

```
In[6]:=
        y[20]
```

```
Out[6]=
        0.15 hp[1] + 0.15 hp[1.15] + 0.15 hp[1.3] +
            0.15 hp[1.45] + 0.15 hp[1.6] +
            0.15 hp[1.75] + 0.15 hp[1.9] +
            0.15 hp[2.05] + 0.15 hp[2.2] +
            0.15 hp[2.35] + 0.15 hp[2.5] +
            0.15 hp[2.65] + 0.15 hp[2.8] +
            0.15 hp[2.95] + 0.15 hp[3.1] +
            0.15 hp[3.25] + 0.15 hp[3.4] +
            0.15 hp[3.55] + 0.15 hp[3.7] + 0.15 hp[3.85]
```

Construct a **Sum** command which will produce the same output and enter it with the definition of **hp**.

```
In[7]:=
        hp[x_]:= 1/x
        Sum[0.15*hp[1 + 0.15*k], {k,0,19}]
```

```
Out[7]=
        1.4443
```

There is the approximation, $h(4) \approx 1.4443$.

Practice Exercise 1: Using $\Delta x = 0.01$, approximate $h(3)$ using a **Sum** command. (*Ans.* 1.10195)

Infinite Sums and Antiderivatives

In general, the procedure goes like this: You want to find $f(b)$ and you know $f'(x)$ and the numerical value of $f(a)$. You decide to use $y(n)$ for the approximation to $f(b)$. You go through the routine above and find that **Sum[fp[x[k]]*dx, {k, 0, n–1}]** is the command that you need. Thus,

$$f(b) \approx f(a) + \sum_{k=0}^{n-1} f'(x_k) \, dx, \qquad \text{where } dx = (b-a)/n.$$

We know that this approximation gets better as the value of n increases and the value of dx decreases. The limit as n approaches infinity of sums of the type that appears on the right side of the equation are symbolized in mathematics with integral signs, so the equation above can be rewritten as

$$f(b) - f(a) = \int_a^b f'(x) \, dx$$

This result is known formally as the Fundamental Theorem of Calculus.

Code for the Practice Exercises

```
1. hp[x_]:= 1/x
   Sum[0.01*hp[1 + 0.01*k], {k,0,199}]
```

Problems

1. Suppose $h'(x) = 1/x$ and $h(1) = 0$. Find $h(2)$
   (a) using the **Sum** technique outlined in this exploration. Apply it several times using $n = 2, 4, 8,\ldots$, until you find at least two results which agree on the first two decimal places;
   (b) using antidifferentiation.
2. Suppose $g'(x) = x^3$ and $g(0) = 1$.
   (a) Calculate $g(2)$ using antidifferentiation.
   (b) Calculate $g(2)$ to two places using a **Sum** command.

20

Numerical Integration

A variety of techniques for performing integration are studied in calculus courses: integration by parts, by substitution techniques, by partial fractions, and so forth. But the fact is that for many functions it is simply not possible to find an antiderivative in terms of familiar functions. And one does not have to go very far afield to find examples of this phenomenon. The integrand for the elliptical arc length calculation studied in Exploration Eighteen, for example, does not have an elementary antiderivative. In such a case, it is not be possible to calculate the definite integral in the usual manner. We need a way of evaluating these integrals that does not require us to find an antiderivative. Thus we are forced to use numerical methods.

This task is not as difficult as it might sound, since a definite integral is defined in terms of an approximating sum (usually referred to as a "Riemann sum," after the mathematician who first formulated the integral in this fashion).

$$\int_a^b f(x)\, dx \approx \sum_{i=1}^n f(z_i)\, h$$

It is understood that the sum approaches the value of the integral as n gets large and h gets small.

Recall the method for setting up the approximating sum: The interval of integration $a \le x \le b$ is divided into n subintervals of equal length. The length of one subinterval is

$h = (b - a)/n$ and the subdivision points are $x_0 = a$, $x_1 = a + h$, $x_2 = a + 2h$, $x_3 = a + 3h$,...
, $x_n = a + nh = b$.

When the subdivision is complete, a point z is chosen from each subinterval. The one selected from the ith subinterval is called z_i.

The next step is to evaluate the function $f(x)$ at the z-numbers and to multiply each of these function values by h. The products are then totaled to form the Riemann sum

$$\sum_{i=1}^{n} f(z_i) \, h.$$

Some Examples of Riemann Sums

For practice, let's construct some Riemann sums for the function $f(x) = 1/x$, on the interval $1 \le x \le 2$, with the interval divided up into five equal parts.

THE LEFT RIEMANN SUM

The left Riemann sum is formed by taking the number z_i to be the left-hand endpoint of the subinterval, in other words, the number x_{i-1}. The sum can be visualized as the total of the area in these five rectangles.

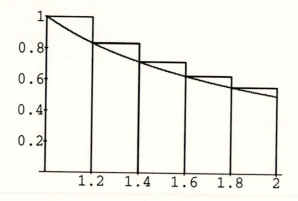

Calculate the left-hand Riemann sum.

```
In[1]:=
        Clear[f,x];
        f[x_]:= 1/x;
        h = (2-1)/5;
        x[k_] := 1 + k*h;
        N[Sum[f[x[k]]*h, {k,0,4}]]
```

Out [1] =
>   0.745635

This result tells us that

$$\int_{1}^{2} 1/x \, dx \approx 0.745635.$$

We can tell from the picture that the approximation is too large.

A RIGHT RIEMANN SUM

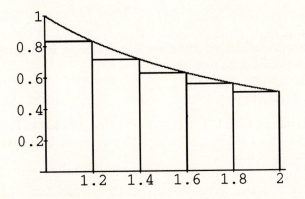

In forming the right Riemann Sum z_i is set equal to x_i, the right-hand endpoint of the subinterval. A small adjustment in the previous **Sum** command calculates this total. It will be smaller than the value of the integral. (Why?)

In[2]:=
>   N[Sum[f[x[k]]*h, {k,1,5}]]

Out[2] =
>   0.645635

A MIDPOINT RIEMANN SUM

Here is the picture:

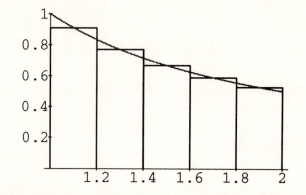

The rectangles are partly above the curve, and partly below it. The errors will tend to cancel out. The midpoint sum, therefore, will more accurately approximate the integral than either the left sums or the right sums.

And here is the total:

```
In[3]:=
        N[Sum[f[(x[k] + x[k+1])/2]*h, {k,0,4}]]

Out[3]=
        0.691908
```

Summing Procedures

We'd like to be able to calculate Riemann sums like these for other functions and for finer subdivisions of the interval of integration, so a procedure which programs *Mathematica* to calculate them automatically is the way to go. The three functions below will do the trick. If you wish, you can copy them into your notebook from the input cell file on your *Exploring Calculus* disk or, if you prefer you can type them in by hand. They are straightforward generalizations of the sum calculations we made above for $1/x$. The only new twist (it won't be new to those who studied the optional section in the arc length exploration) is the **Block** command. It turns the variables h, x, and i into *local* variables which means that it automatically clears them of any previous assignment before the function is executed. Then, when the execution is complete, it restores the former values.

```
In[4]:=
        leftRiemannSum[f_,a_,b_,n_]:=
            Block[{h,x,i},
                h = (b-a)/n;
                x[i_]:=a + h*i;
                N[Sum[f[x[i]]*h, {i,0,n-1}]]
                ];
```

```
rightRiemannSum[f_,a_,b_,n_]:=
    Block[{h,x,i},
        h = (b-a)/n;
        x[i_]:=a + h*i;
        N[Sum[f[x[i]]*h,  {i,1,n}]]
        ];

midPointSum[f_,a_,b_,n_]:=
    Block[{h,x,i},
        h = (b-a)/n;
        x[i_]:=a + h*i + h/2;
        N[Sum[f[x[i]]*h,  {i,0,n-1}]]
        ];
```

Error Analysis

When approximating an integral it is important to select a technique which minimizes the error without using an extravagant amount of computer time to do it. We're going to try to get a handle on the sort of errors which these three methods manifest. First, we need a table of approximations made by the three types of sums. We will set it up so that it compares the right, left, and midpoint sums for $n = 5$, 10, and 20.

```
In[5]:=
    Clear[n,k];
    n = 5*2^k;
    approx = Table[
        {n,
         leftRiemannSum[f,1,2,n],
         rightRiemannSum[f,1,2,n],
         midPointSum[f,1,2,n]
        },
        {k,0,2}
        ];
    TableForm[
        approx,
        TableHeadings->{
          None,
          {"n\n","LSum\n","RSum\n","MPSum\n"}
          }
        ]
```

```
Out[5]//TableForm=
        n      LSum         RSum         MPSum

        5      0.745635     0.645635     0.691908
        10     0.718771     0.668771     0.692835
        20     0.705803     0.680803     0.693069
```

We know that the left sums are too large, the right sums are too small, and that the midpoint sums are the most accurate for each number of subdivisions, but we do not yet know the magnitude of the errors. We can figure them out, because for this function the antiderivative *is* known.

In[6]:=

```
      Integrate[1/x,x]
```

Out[6]=

```
      Log[x]
```

Now, compute the *difference* in the value of the antiderivative at 2 and its value at 1. Get eight digits while you are at it and assign the number the name **int**.

In[7]:=

```
      int = N[Log[2] - Log[1], 8]
```

Out[7]=

```
      0.69314718
```

Next, we edit our previous table and transform it into a table of errors by making each entry the difference between the approximation and the number **int**. The **TableHeadings** option must also be changed.

In[8]:=

```
      Clear[n,k];
      n = 5*2^k;
      errors = Table[
          {n,
          leftRiemannSum[f,1,2,n] - int,
          rightRiemannSum[f,1,2,n] - int,
          midPointSum[f,1,2,n] - int
          },
          {k,0,2}
          ];
```

```
TableForm[
    errors,
    TableHeadings->
      {None,
      {"n\n",
      "LSum\nerror\n",
      "RSum\nerror\n",
      "MPSum\nerror\n"}
      }
    ]
```

Out[8]//TableForm=

n	LSum error	RSum error	MPSum error
5	0.0524877	-0.0475123	-0.00123929
10	0.0256242	-0.0243758	-0.00031182
20	0.0126562	-0.0123438	-0.0000780823

Some interesting observations can be made from this error table. Notice that for each of the approximation techniques, the error gets smaller as n increases. The left Riemann sum and the right Riemann sum have opposite signs, as we saw from the graph, and the magnitude of the left sums is about the same as that of the right sums. The midpoint errors, on the other hand are much smaller for all three values of n.

What patterns do you see as you read down a column? For the left Riemann sum (second column) the error seems to be cut about in half from one row to the next. It seems that if you double the number of intervals the error in the left Riemann sum is divided by 2. The same pattern holds for the right Riemann sum. What pattern do you see the midpoint Riemann sum?

Practice Exercise 1: Modify the preceding commands to give two more rows in the errors table, for $n = 40$ and $n = 80$. Do the patterns we observed in the errors still hold?

Trapezoidal Sums

Since the left Riemann sums are too large and the right Riemann sums too small, averaging the two might cancel out much of the error. Perhaps such an approximation would have even smaller errors than the midpoint sums. It's worth a try.

```
In[9]:=
    trapezoidalSum[f_,a_,b_,n_]:=
      (leftRiemannSum[f,a,b,n]+
       rightRiemannSum[f,a,b,n])/2
```

Create **approx** and **errors** for the trapezoidal sums. See how it fares in comparison to the midpoint sums. Once again you can edit the previous code for these two tables.

```
In[10]:=
        Clear[n,k];
        n = 5*2^k;
        approx = Table[
            {n,
             trapezoidalSum[f,1,2,n],
             midPointSum[f,1,2,n]
             },
             {k,0,2}
             ];
        TableForm[
            approx,
            TableHeadings->
              {None,
               {"n\n","TrapSum\n","MPSum\n"}}
             ]
```

```
Out[10]//TableForm=
        n     TrapSum     MPSum

        5     0.695635    0.691908
        10    0.693771    0.692835
        20    0.693303    0.693069
```

```
In[11]:=
        Clear[n,k];
        n = 5*2^k;
        errors = Table[
           {n,
            trapezoidalSum[f,1,2,n]-int,
            midPointSum[f,1,2,n]-int
            },
            {k,0,2}];
        TableForm[
           errors,
           TableHeadings->
             {None,
              {"n\n","TrapSum\nerror",
                "MPSum\nerror\n"}
              }
            ]
```

```
Out[11]//TableForm=
        n       TrapSum         MPSum
                error           error

        5       0.00248774      -0.00123929
        10      0.000624223     -0.00031182
        20      0.000156201     -0.0000780823
```

The magnitude of the trapezoidal error appears to be about twice that of the midpoint sums, but with the opposite sign. Its error also seems to be decreasing by a factor of 4 when *n* is doubled.

Practice Exercise 2: Modify the preceding commands to give two more rows in the **errors** table. Do the patterns we observed still hold?

Most calculus books include a section on trapezoidal sums. Usually it is entitled the Trapezoidal Rule and includes a formula for calculating trapezoidal totals that does not make direct reference to right or left sums. For example, according to the Trapezoidal Rule taken from a standard calculus text, the trapezoidal approximation for $n = 5$ would be calculated as follows:

```
(1. f[1.] + 2. f[1.2] + 2. f[1.4] +
    2. f[1.6] + 2. f[1.8] + 1. f[2.]) / 10
```

Notice that the sum is written explicitly in terms of function values. As it turns out, our **trapezoidalSum** is equivalent. To see this, **Clear** the definition of **f** and ask *Mathematica* for **trapezoidalSum**. It will return a result in terms of *f* because it will no longer know a rule for the function. Then we can see if the *Mathematica* output squares with the standard rule.

```
In[12]:=
    Clear[f];
    n = 5;
    a = 1;
    b = 2;
    trapezoidalSum[f,1,2,n]

Out[12]=
    (0.2 f[1.] + 0.4 f[1.2] + 0.4 f[1.4] +
        0.4 f[1.6] + 0.4 f[1.8] + 0.2 f[2.]) / 2
```

That is the calculation *Mathematica* would have made had it known a rule for *f*. Sure enough it is the equivalent to the expression yielded by the Trapezoidal Rule. Multiplying top and bottom by 5 produces that formula.

```
In[13]:=
    Expand[5*Numerator[%]]/
        (5*Denominator[%])
```

```
Out[13]=
        (1. f[1.] + 2. f[1.2] + 2. f[1.4] +
          2. f[1.6] + 2. f[1.8] + 1. f[2.]) / 10
```

Simpson's Rule

This is another stab at getting a method which improves midpoint sums: The error in the midpoint sum and that from trapezoidal sum are opposite in sign, but the trapezoidal error is approximately twice as large. Perhaps we could cancel out the errors by taking

$$\frac{2 * (\text{Midpoint Sum}) + \text{Trapezoid Sum}}{3}$$

Notice that the numerator contains three approximations in all. Division by 3 computes their average.

Define a new function which does this and set up a table to compare the errors.

```
In[14]:=
        simpsonSum[f_,a_,b_,n_]:=
          (2*midPointSum[f,a,b,n]+
            trapezoidalSum[f,a,b,n])/3


In[15]:=
        Clear[n,k];
        n = 5*2^k;
        f[x_]:= 1/x;
        errors = Table[
            {n,
            trapezoidalSum[f,1,2,n] - int,
            simpsonSum[f,1,2,n] - int
            },
            {k,0,2}
            ];
        TableForm[
            errors,
            TableHeadings->
              {None,
              {"n\n","TrapSum\nerror",
              "simpSum\nerror\n"}
              }
            ]
```

```
Out[15]//TableForm=
        n      TrapSum         simpSum
               error           error
        5      0.00248774      3.05013 10 -6
        10     0.000624223     1.94105 10 -7
        20     0.000156201     1.2188 10 -8
```

Simpson sums are the hands down winners. Let's see what the sum is for $n = 20$.

In[16]:=

```
        simpsonSum[f,1,2,20]
```

Out[16]=

```
        0.693147
```

As with trapezoidal sums, Simpson sums are also treated in the standard calculus text, under the heading Simpson's Rule. The rule is generally stated without reference to other summing techniques and, instead, gives a calculation formula directly in terms of function values. Our **simpsonSum** and Simpson's Rule are equivalent and this fact can be established in a manner similar to our treatment of trapezoidal equivalence.

Of the methods of numerical integration presented in this exploration, Simpson's Rule is the best. It has much greater accuracy than the other methods, at the price of only somewhat more complicated code. However, for serious numerical integration, other algorithms are known which give even greater accuracy or are better able to deal with badly behaved functions. If you are interested in this topic, see a book on numerical analysis.

Problems

1. Consider the integral of $f(x) = x^2$ from $x = 1$ to $x = 3$. This function is increasing on the interval $1 \le x \le 3$. Make a table of the left Riemann sum, the right Riemann sum, the Trapezoid Rule, and the midpoint sum. Use $n = 5, 10$, and 20. Make a table of the errors. Does the same pattern in the errors we found for $f(x) = 1/x$ hold for $f(x) = x^2$?

2. Use Simpson's Rule with $n = 2$ to approximate the following integrals. Find the error. Comment on your observations.

   (a) $\displaystyle\int_{2}^{7} x \, dx$

   (b) $\displaystyle\int_{3}^{8} x^2 \, dx$

(c) $\displaystyle\int_{1}^{6} x^3\ dx$

(d) $\displaystyle\int_{2}^{4} (5x^2 - 3x + 2)\ dx$

(e) $\displaystyle\int_{-1}^{2} x^4\ dx$

3. Let $f(x) = 1/x$ and consider the integral of f from $x = 1$ to $x = 3$. Use $n = 3, 6$, and 12 to find the Simpson's Rule approximation. Find the error each time, and try to guess how the error changes as n is doubled. Test your guess by finding the error when $n = 24$.

21

The Exponential Function and *e*

The number *e* turns up in many different mathematical contexts. Along with 0, 1, and π, it is considered one of the four most important numbers in mathematics. In this project we will investigate the number *e* and its use as the base for exponential functions.

The Number *e*

Suppose that you find an investment that pays 100% interest per year. That is, for each dollar you invest, you will have $2 at the end of one year. This is already a *very* good deal; but if you can get the interest compounded more than once a year, you will do even better. For example, if the interest is compounded twice a year, you will be paid 50% interest after six months, so each dollar you invested will be worth $1.50. Then you will get another 50% interest, paid on the $1.50, at the end of the year. This gives $1.50 * (1.5) = $2.25, an extra $0.25 on each dollar.

Compounding more frequently will give you an even greater return. Say the compounding is four times per year. Then each dollar will be multiplied by 1.25 at the end of the first quarter, again by 1.25 at the end of the second quarter, by a third 1.25 at the end of the third quarter, and, finally by 1.25 again at the end of the year. That is, each dollar is multiplied by a total of $(1.25)^4$ in one year.

```
In[1]:=
        (1.25)^4

Out[1]=
        2.44141
```

Practice Exercise 1: How much do you get after one year at 100% interest compounded 12 times (monthly)? (*Ans.* $2.61)

If the interest is compounded *n* times per year, each dollar will be multiplied by $(1 + 1/n)^n$ by the end of the year. Define this multiplier as a function of *n*.

```
In[2]:= Clear[f,n]
        f[n_]:= (1.0 + 1.0/n)^n
```

Make a table of values for $f(n)$.

```
In[3]:=
        Clear[t,n];
        TableForm[
          Table[{n,f[n]}, {n,1,10}],
          TableHeadings->{None, {"n\n"," f(n)\n"}}
          ]

Out[3]//TableForm=
          n      f(n)

          1      2.
          2      2.25
          3      2.37037
          4      2.44141
          5      2.48832
          6      2.52163
          7      2.5465
          8      2.56578
          9      2.58117
          10     2.59374
```

From the first line in the table, we see that when 100% interest is paid once a year, the amount is doubled in that year. When the interest is compounded twice a year ($n = 2$) each dollar grows to $2.25 in one year, just as we saw earlier. And when the interest is compounded quarterly ($n = 4$), each dollar becomes $2.44, again, the same number we calculated earlier.

Practice Exercise 2: Use f[n] to find the multiplier for monthly compounding; daily compounding. How much more do you get per dollar when compounding is done every day compared to every month? (*Ans.* About 10.1 cents)

What happens as the compounding is done more and more often? Make a table of values for large n, by taking $n = 10^k$.

```
In[4]:=
      Clear[t,k];
      TableForm[
        Table[{10.^k,  f[10.^k]},  {k,1,8}],
        TableHeadings->{None,  {"n\n","f(n)\n"}}
        ]
```

```
Out[4]//TableForm=
            n              f(n)

          10.            2.59374
          100.           2.70481
          1000.          2.71692
          10000.         2.71815
          100000.        2.71827
          1. 10⁶         2.71828
          1. 10⁷         2.71828
          1. 10⁸         2.71828
```

As n gets larger, $f(n)$ also gets larger; but its values do not appear to get infinitely large. In fact, it appears that $f(n)$ approaches 2.71828... as $n \to \infty$. The number e is defined to be this limit, so it looks like $e = 2.71828...$. Restated in terms of interest, this means that as 100% interest is compounded more and more often, the amount you have at the end of one year per dollar invested gets closer and closer to e.

If the original amount is simply multiplied by e at the end of the year, we say that the interest is compounded *continuously*. The effect is as if the interest were compounded an infinite number of times in the year.

The constant e is built in to *Mathematica*. Here is its 10-digit value:

```
In[5]:=
      N[E,10]
```

```
Out[5]=
       2.718281828
```

The Exponential Function

When an interest rate of 100% is compounded continuously, each dollar is multiplied by $e = 2.71828...$ in a year. So in two years, each dollar would be multiplied by e for the first year, and then by e again for the second year. That is, it would be multiplied by e^2. Similarly, in three years each dollar would be multiplied by e^3. In general, in t years each dollar would be multiplied by e^t. Therefore, e^t gives the amount to which $1 grows when invested at 100% per year compounded continuously for *t* years.

Example: You invest $300 at 100% per year compounded continuously. How much will you have in four years?

```
In[6]:=
        N[300*E^4]
```

```
Out[6]=
        16379.4
```

Wow! You would have more than $16,000!

Practice Exercise 3: Suppose $400 is invested at 100% interest. How much is it worth at the end of six years? (*Ans.* $161,372!)

The values of e^t certainly go up very quickly. In the next two sections we'll take a closer look at this function and, in particular, try to get some measure of exactly how fast it is rising.

THE GRAPH OF e^t

We can get a rough idea of how this function behaves with a quick plot, say for $-2 \le t \le 4$.

```
In[7]:=
        Plot[E^t,  {t,-2,4}];
```

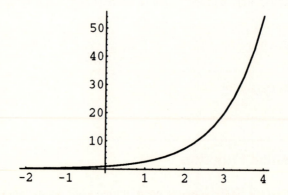

It looks like the curve is rising and concave up, which means that as $t \to +\infty$, the values of e^t increase at a more and more rapid rate. That's pretty much what we figured. The plot also suggests that as $t \to -\infty$, e^t gets smaller and smaller. Check it out.

```
In[8]:=
        Clear[t];
        Limit[E^t, t->-Infinity]
```

General::load: Loading package Series'.

```
Out[8]=
        0
```

The *x*-axis is a horizontal asymptote.

So, we have a rough picture of how e^t changes with t. As t goes from $-\infty$ to $+\infty$, e^t increases in value from infinitely close to zero to infinitely large, and it increases at a faster and faster rate. It just zooms up. Good. Now let's see if we can make these ideas a little more precise. For example, how fast is fast? How long does it take e^t to double in value?

CHARACTERISTIC DOUBLING TIME

We have seen that each dollar invested at 100% interest per year compounded continuously will be multiplied by 2.71828 in one year. How long does it take for the money to exactly double? Clearly less than a year, but exactly how long? Here's a clever way to get at this question: Plot the function e^t and the horizontal line at 2.

```
In[9]:=
        Plot[
          {2,E^t}, {t,0,1},
          AxesLabel->{"t", "E^t"}
          ];
```

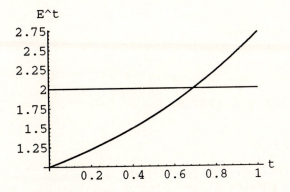

When $t = .7$ (about), the value of e^t is 2. Now that we know approximately how long it takes, we can use FindRoot to get a more accurate reading.

```
In[10]:=
        FindRoot[E^t==2, {t,0.7}]
```

```
Out[10]=
        {t -> 0.693147}
```

The investment exactly doubles in 0.693 years (about 33 weeks).

How long does it take to double again? We want the *t* value for which $e^t = 4$. Again estimating from the graph:

```
In[11]:=
        Plot[
          {4,E^t},{t,0,2},
          AxesLabel->{"t",  "E^t"}
          ];
```

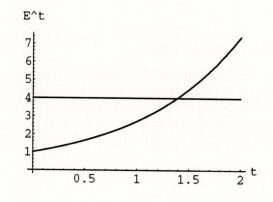

```
In[12]:=
        FindRoot[E^t==4,   {t,1.5}]
```

```
Out[12]=
        {t -> 1.38629}
```

So, it takes

```
In[13]:=
        1.38629 - 0.693147
```

```
Out[13]=
        0.693143
```

Note that this is almost exactly the same length of time it took to double from *t* = 0! Will it double again, grow to 8, in another 0.693147 years? Let's check. The time will be *t* = 1.38629 + 0.693147.

```
In[14]:=
        1.38629 + 0.693147
```

```
Out[14]=
        2.07944
```

```
In[15]:=
        N[E^2.07944]
```

```
Out[15]=
        7.99999
```

That looks a lot like 8. It turns out that no matter what value of t you start with, the value of e^t is doubled when t is increased by 0.693143. Better test this wild claim.

Practice Exercise 4: Find the times at which e^t equals 3, 6, 12, and 24. Are the intervals between these times equal to 0.693143? (*Ans.* Yes)

The time interval $t = 0.693143$ over which e^t will always double in value is called its *characteristic* doubling time. Most functions have no such property. Think about the behavior of t^2 for example.

```
In[16]:=
        N[Solve[ t^2==1, t]]
```

```
Out[16]=
        {{t -> 1.}, {t -> -1.}}
```

```
In[17]:=N[Solve[ t^2==2, t]]
```

```
Out[17]=
        {{t -> 1.41421}, {t -> -1.41421}}
```

```
In[18]:=
        N[Solve[ t^2==4, t]]
```

```
Out[18]=
        {{t -> 2.}, {t -> -2.}}
```

```
In[19]:=
        N[Solve[ t^2==8, t]]
```

```
Out[19]=
        {{t -> 2.82843}, {t -> -2.82843}}
```

```
In[20]:=
        N[Solve[ t^2==16, t]]
```

```
Out[20]=
        {{t -> 4.}, {t -> -4.}}
```

Compare the positive time values, and you quickly see that the interval between one time value and the next is different. The function t^2 does not have a characteristic doubling time.

Practice Exercise 5: Show that 10^t has a characteristic doubling time and find its value. (*Ans.* 0.30103)

How Fast Does an Exponential Function Grow?

As $t \to \infty$, the function e^t grows very quickly; there are other functions that grow quickly also. How do their rates of growth compare? Consider t^{10}. Plot e^t and t^{10} on the same axis, for $0 \le t \le 2$.

```
In[21]:=
        Plot[
          {E^t,t^10},  {t,0,2},
          PlotStyle->{{Dashing[{.02}]},{}}
          ];
```

The exponential is the dashed curve. It seems pretty clear from this graph that t^{10} grows faster than e^t. But wait! We may not be looking at large enough values for t. Try plotting again, this time with the t interval 0 to 40.

```
In[22]:=
        Plot[
          {E^t,  t^10},  {t,0,40},
          PlotStyle->{  {Dashing[{.02}]},{}  }
          ];
```

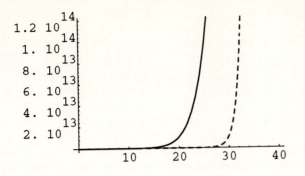

The exponential is still below t^{10}, but this time it looks like it may be catching up. It is hard to tell because the vertical range is not tall enough to show how the curves are behaving between $t = 30$ and $t = 40$. Try plotting for that interval.

```
In[23]:=
    Plot[
       {E^t, t^10}, {t,30,40},
       PlotStyle->{ {Dashing[{.02}]},{} }
       ];
```

Now you can see the exponential curve pulling ahead. Where do the curves cross? A table will be useful in pinning down this crossing point.

```
In[24]:=
    TableForm[
       Table[{t,N[E^t],t^10}, {t,33,38,1.}],
       TableHeadings->
          {None, {"t\n","E^t\n","t^10\n"} }
       ]
```

```
Out[24]//TableForm=
          t        E^t              t^10

          33      2.14644 10⁴        1531578985264449
          34.     5.83462 10⁴       2.06438 10⁵
          35.     1.58601 10⁵       2.75855 10⁵
          36.     4.31123 10⁵       3.65616 10⁵
          37.     1.17191 10⁶       4.80858 10⁵
          38.     3.18559 10⁶       6.27821 10⁵
```

The crossover is between $t = 35$ and $t = 36$. So, after $t = 36$, the function e^t is ahead. Does it stay ahead? Have *Mathematica* calculate a table of values for the ratio of t^{10} to e^t as t gets large:

```
In[25]:=
        Clear[t];
        t = 10^k;
        TableForm[
          Table[{t,N[t^10/E^t]}, {k,1,6}],
          TableHeadings->
            {None, {"t\n","t^10/E^t\n"}}
          ]

Out[25]//TableForm=
          t        t^10/E^t

          10       453999.
          100      3.72008 10²⁴
          1000     5.07596 10⁴⁰⁵
          10000    1.13548 10⁴³⁰³
          100000   3.5629495653094 10⁴³³⁸⁰
          1000000  3.296831478089 10⁴³⁴²³⁵
```

The ratio becomes exceedingly small. That means that e^t grows much faster than t^{10} for large enough t. This argument by itself doesn't eliminate the possibility of another crossing, but in fact there is none. The function e^t does grow faster than t^{10} when t gets large enough.

Code for Practice Exercises

1. `N[(1 + (1/12))^12]`
2. `N[(1 + (1/365))^365 - (1 + (1/12))^12]`
3. `N[400*E^6]`

```
4. r1 = FindRoot[E^t==3,      {t, 1.1}]
   r2 = FindRoot[E^t==6,      {t, 1.7}]
   r3 = FindRoot[E^t==12,     {t, 2.4}]
   r4 = FindRoot[E^t==24,     {t, 3.1}]
   (t/.r2) - (t/.r1)
   (t/.r3) - (t/.r2)
   (t/.r4) - (t/.r3)
5. r1 = FindRoot[10^t==2,     {t, 0.3}]
   r2 = FindRoot[10^t==4,     {t, 0.6}]
   r3 = FindRoot[10^t==8,     {t, 0.9}]
   r4 = FindRoot[10^t==16,    {t, 1.2}]
   (t/.r2) - (t/.r1)
   (t/.r3) - (t/.r2)
   (t/.r4) - (t/.r3)
```

Problems

1. Use the built-in function E to find the value of e to 14 decimal places.

2. Suppose that money is invested at 100% interest, compounded every hour. How much does a dollar grow to in a year?

3. Suppose that money is invested at 100% interest, compounded every second. How much does a dollar grow to in a year?

4. How much more do you get per dollar if 100% interest is compounded every second compared to its being compounded every month? (You may need to use N to get more decimal places than the default.)

5. How much more do you get per dollar if 100% interest is compounded continuously compared to its being compounded every day?

6. Estimate the time it takes e^t to triple, starting from $t = 0$. Does it triple again in the same length of time? (That is, is there a characteristic tripling time?)

7. How long does it take the exponential e^t to double, starting at $t = 1$?

8. Estimate the value k so that for $t > k$ it is true that $e^t > t^4$.

22

Exponential Decay

In this exploration you will be using *Mathematica* to analyze data gathered from the decay of a radioactive substance. The goal is to determine a formula which models the decay and to find the half-life of the substance. Before getting into the specific details of the radioactive decay data, take the time to review some helpful mathematical ideas and their related *Mathematica* commands. We begin with a problem which helps to set the stage for the discussion.

Linear Decay

A student who doesn't trust banks has $10,000 in cash stored in a mattress. Every semester for four years the student removes $1000 to help pay school expenses. Since a total of $2000 is spent each year, the amount of money a in the mattress at the end of x school years is given by $a(x) = 10,000 - 2000 x$. Teach *Mathematica* the formula and generate a table which shows the money remaining in the mattress at the end of each year, call the table **amts1**. (The last character in the name is the number one it is *not* the letter "l".)

```
In[1]:=
        Clear[a,x]
        a[x_]:= 10000 - 2000x
        amts1 = Table[{x,a[x]},{x,0,4}]
```

```
Out[1]=
        {{0, 10000}, {1, 8000}, {2, 6000}, {3, 4000},
          {4, 2000}}
```

As you may recall, *Mathematica* has a variety of commands which help get information from tables. For example, you can determine the number of elements in a table, a quantity referred to in mathematics as the *length* of the table.

```
In[2]:=
        Length[amts1]
```

```
Out[2]=
        5
```

Look at the table vertically

```
In[3]:=
        TableForm[amts1]
```

```
Out[3]//TableForm=
        0    10000
        1    8000
        2    6000
        3    4000
        4    2000
```

When the table is written this way, it is clear that it has 5 rows and 2 columns. Individual elements in the table are identified by their row and column positions. For example, the number 4000 is in the third row and the second column. If you type in **amts1[[4, 2]]**, *Mathematica* will return 4000.

```
In[4]:=
        amts1[[4,2]]
```

```
Out[4]=
        4000
```

Here is the third row:

```
In[5]:=
        Table[amts1[[3,j]],{j,1,2}]
```

```
Out[5]=
        {2, 6000}
```

And the first column:

```
In[6]:=
        Table[amts1[[i,1]],{i,1,5}]//TableForm
```

```
Out[6]//TableForm=
        0
        1
        2
        3
        4
```

A part of a table can be transformed. For example, the next command rewrites the table **amts1** so that the second entry in each row gives the number of thousands of dollars remaining each year rather than the number of dollars.

```
In[7]:=
        Table[
            {amts1[[i,1]],  (amts1[[i,2]])/1000},
            {i,1,Length[amts1]}
            ]//TableForm
```

```
Out[7]//TableForm=
        0    10
        1    8
        2    6
        3    4
        4    2
```

Practice Exercise 1: Write a command which yields the first three elements of the second column of **amts1**.

Practice Exercise 2: Write a command which gives the **amts1** table with the first column entries squared.

Let's graph the **amts1** table. The plot is easier to see with an enlarged **PointSize** and a **PlotRange** that includes the origin and runs a little above 10,000 .

```
In[8]:=
        ListPlot[
           amts1,
           PlotStyle->PointSize[.03],
           PlotRange->{0,11000}
           ];
```

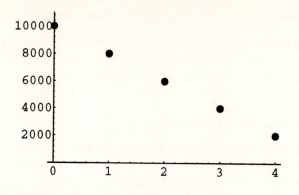

Since the money in the mattress decreases by the same amount each year, the points fall on a straight line with a slope of –2000 and a y-intercept of 10,000. This line and the plot of the table can be graphed simultaneously.

```
In[9]:=
        plot1 = Plot[
           10000-2000x,  {x,0,4},
           Epilog->{{
             PointSize[0.03],
             Map[Point,amts1]
             }},
           PlotRange->{0,11000}
           ];
```

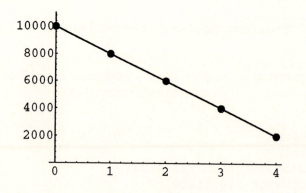

The money in the mattress is an example of a substance that decays linearly.

Exponential Decay

A second student equally distrustful of the thrifts also has $10,000 stored in a mattress, but she has a different scheme for withdrawing it. This student removes 20% of the money in the mattress at the beginning of each school year, so the amount of money she has at the end of a school year will be 80% of the previous year's amount. Here are the calculations for the four-year period.

After the first year the amount remaining is

```
In[10]:=
        10000(.80)

Out[10]=
        8000.
```

Practice Exercise 3: Calculate the amounts remaining after the second, third, and fourth years. (*Ans.* $6400, $5120, $4096)

The amount of money b which this student has in the mattress at the end of x years is predicted by $b(x) = 10000 \, (.80)^x$. Enter this formula and generate a table of the yearly amounts called **amts2** and plot it in gray points.

```
In[11]:=
        b[x_]:= 10000.0(.80)^x;
        amts2 = Table[{x,b[x]},{x,0,4}];
        plot2 = ListPlot[
            amts2,
            PlotStyle->{
                PointSize[.03],
                GrayLevel[.5]
                },
            PlotRange->{0,11000}
            ];
```

Plot the curve $b(x)$ and show it superimposed on the plot of **amts2**.

```
In[12]:=
        plot3 = Plot[
            b[x],{x,0,4},
            Prolog->{{
              GrayLevel[.5],
              PointSize[.03],
              Map[Point,amts2]
              }},
            PlotRange->{0,11000}
            ];
```

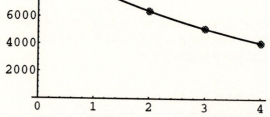

This is an example of *exponential* decay. The points fall on the exponential curve $10000 (.80)^x$. Compare the plots for exponential decay and linear decay.

```
In[13]:=
        Show[{plot1,plot3,plot2}];
```

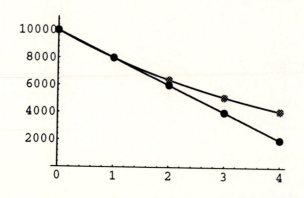

In the case of exponential decay, the size of the drop in the amount each year diminishes so that, except for the first two points, the plots do not coincide.

Here is an interesting characteristic of the exponential decay: although the amounts do not decrease linearly, the logarithms of the amounts do.

In[14]:=
```
ListPlot[
    Table[{x,Log[b[x]]},{x,0,4}],
    PlotStyle->PointSize[.03],
    AxesLabel->{"x","Log[b[x]]"}
    ];
```

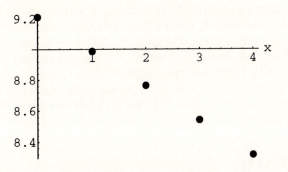

It's easy to see why these points fall in a straight line:

$$Log(b(x)) = Log(10000(.80)^x)) = Log(10000) + x\,Log(0.80).$$

Both the quantities, Log(10000) and Log(.80) are constants.

In[15]:=
```
{Log[10000.],  Log[.80]}
```

Out[15]=
```
{9.21034,  -0.223144}
```

So Log($b(x)$) = 9.21034 − 0.21034 x, which is a linear function of x.

Shortly, we will use some of the notions developed here in the study of radioactive decay data. Meanwhile, we have one more preliminary topic to look at. Before moving on to it, **Clear** your variables.

In[16]:=
```
Clear[a,b,amts1,amts2,plot1,plot2,plot3]
```

Exponential Curve Fitting

Imagine that you have collected a list of eight data points from an experiment.

$$\{1, 4.0\}, \{2, 3.3\}, \{3, 2.8\}, \{4, 2.3\}, \{5, 1.8\}, \{6, 1.6\}, \{7, 1.1\}, \{8, .9\}$$

Further, let's say that you have reason to believe that points can be fit by an exponential curve. The task is to find the curve.

Begin by assigning the list the name **data**. The most straightforward way to do this in to type all the pairs in by hand and enter the list.

```
data = {{1,4.0},{2,3.3},{3,2.8},
     {4,2.3},{5,1.8},{6,1.6},{7,1.1},{8,.9}}
```

In this example the list is short, so such a technique is feasible. In real life it would not be since a data set could be much, much longer. *Mathematica* has a command called **ReadList** which allows you to organize raw data from a file into your notebook and assign the list a name. For example, the 16 numbers in the eight sample data points on the preceding page are contained in a file called **samp.data** which has been placed in the **data** file on your *Exploring Calculus* disk. Here is how you ask *Mathematica* to organize those 16 numbers into a list of pairs called **data** and place the list in your notebook. (You will have to find **samp.data** for *Mathematica* just as you do with a package.)

```
In[17]:=
     Clear[data];
     data = ReadList[
       "samp.data",
       {Number, Number}
       ]
```

```
Out[17]=
     {{1, 4.}, {2, 3.3}, {3, 2.8}, {4, 2.3},
       {5, 1.8}, {6, 1.6}, {7, 1.1}, {8, 0.9}}
```

Now, see what the data looks like.

```
In[18]:=
     ListPlot[
       data,
       PlotStyle->PointSize[.03],
       PlotRange->{0,4.1}
       ];
```

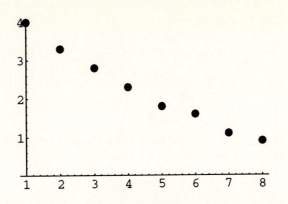

The first thing to check is whether of not it is reasonable to assume that the decay being observed here is exponential. Plot the logs of the second coordinates of the points against the first coordinates and see if the resulting points lie on a line or close to it.

```
In[19]:=
        Clear[k];
        logdata = Table[
           {data[[k,1]], Log[data[[k,2]]]},
           {k,1,Length[data]}
           ]
```

```
Out[19]=
        {{1, 1.38629}, {2, 1.19392}, {3, 1.02962},
         {4, 0.832909}, {5, 0.587787},
         {6, 0.470004}, {7, 0.0953102},
         {8, -0.105361}}
```

The first coordinates of the points in this list match the first coordinates of the points of the list **data**; however, the second coordinates do not. They are the logarithms of the second coordinate numbers in **data**. Plotting this list will be the test. If these points are nearly on a straight line, we can be fairly sure that the points in **data** fall on an exponential curve.

```
In[20]:=
        ListPlot[
          logdata,
          PlotStyle->PointSize[.03],
          AxesLabel->{"t","Log[y]"}
          ];
```

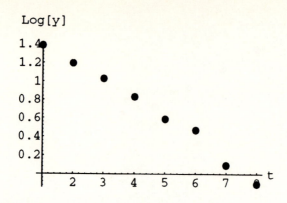

Let *Mathematica* tell you the equation of the line which fits these points best:

```
In[21]:=
        Clear[y];
        logy = Fit[logdata,{1,t},t]
```

```
Out[21]=
        1.64302 - 0.212603 t
```

A note about the command **Fit**: The first part of the command is a list of points like **logdata** in our example, and *Mathematica* returns the function which fits the data points as closely as possible. The last two parts of the command {**1, t**}, **t** tell *Mathematica* that you want the data approximated by an expression of the form $a + bt$, in other words a linear expression. If you had typed {**1, t, t^2**}, **t** instead, *Mathematica* would have returned a quadratic fit in the form $a + bt + ct^2$.

How good is the fit?

```
In[22]:=
        Plot[
           logy,  {t,0,8},
           AxesLabel->{"t","log[y]"},
           Epilog->{{
             PointSize[.03],
             Map[Point,logdata]
             }}
           ];
```

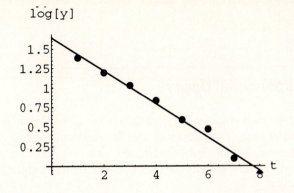

log[y]

Not too shabby. The line fits the data reasonably well. Remember that each of the individual points produced from the data has coordinates of the form $(t, \text{Log}(y))$. Since these points lie very close to the line, we can characterize the relationship between t and $\text{Log}(y)$ with the following equation:

$$\text{Log}(y) = 1.64302 - 0.212603t$$

Therefore, if the data points are falling exponentially, the exponential curve

$$y = e^{1.64302 - 0.212603t}$$

ought to fit them pretty well. This is the acid test.

```
In[23]:=
        Plot[
          E^(1.64302 -0.212603t), {t,0,8},
          Epilog->{{
            PointSize[.03],
            Map[Point,data]
            }},
          PlotRange->{0,5}
          ];
```

Looks good. It seems reasonable to say that the values of *y* decay exponentially as *t* increases.

Half-life for Exponential Decay

In the preceding exploration you saw that quantities which grow exponentially have a characteristic doubling time. The analogous trait for exponential decay is *half-life*. The half-life of a decaying quantity is the time it takes for an amount of the substance to be reduced by half. We'll make the half-life calculation for the exponential function we found in the last section.

```
In[24]:=
        Clear[f,t];
        f[t_]:= E^(1.64302 - 0.212603t)
```

Find out how large *f*(*t*) is initially.

```
In[25]:=
        f[0]
```

```
Out[25]=
        5.17076
```

Now, find the value of *t* when *f*(*t*) is half 5.17076.

```
In[26]:=
        Solve[f[t]==(5.17076 / 2), t]
```

```
Solve::ifun:
            Warning: inverse functions are being used
              by Solve, so some solutions may not be
              found.
```

```
Out[26]=
        {{t -> 3.26029}}
```

If this number really is half-life, then any function value ought to be reduced by half if *t* is increased by 3.26029. Let's run a test. We'll pick a positive number at random, set the exponential expression equal to the number and solve for *t*. Then we'll see if increasing *t* by 3.26029 cuts the function value in half. Begin by asking *Mathematica* to randomly select an integer between, say, 1 and 1000 and call this number *n*. The value you get for **n** will almost certainly be different than the number shown below. (What are the odds?)

```
In[27]:=
        n = Random[Integer, {1,1000}]
```

```
Out[27]=
        997
```

Set $f(t)$ equal to this integer and solve for t.

```
In[28]:=
        Solve[f[t] == n, t]
```

```
Solve::ifun:
            Warning: inverse functions are being used
            by Solve, so some solutions may not be
            found.
```

```
Out[28]=
        {{t -> -24.7491}}
```

Calculate the function value at this value of t plus 3.26029. The value returned should be half of n.

```
In[29]:=
        f[t + 3.26029]/.%
```

```
Out[29]=
        {498.5}
```

It is. Looks like the half-life is 3.26029.

We'd rather be certain. Here's an alternate method for finding the half-life for $f(t)$. See if it gives the same value. The argument begins like this: If h is the half-life for $f(t)$, then $f(t+h))/f(t) = 0.5$ or, equivalently, $\mathrm{Log}(f(t+h)/f(t)) = \mathrm{Log}[.5]$. *Mathematica* can help evaluate the two sides of this log equation.

```
In[30]:=
        Clear[t,x,h];
        Log[f[t + h]/f[t] ]
```

```
Out[30]=
            0. + 0.212603 t - 0.212603 (h + t)
        Log[E                                    ]
```

This Log expression simplifies to

```
        0.212603*t - 0.212603*(h + t)
```

Set it equal to Log[.5] and solve for h.

```
In[31]:=
        Solve[
            0.212603*t-0.212603*(h + t)==Log[.5],
            h]
```

```
Out[31]=
        {{h -> 3.26029}}
```

Looks like we were right on the money the first time. The half-life is definitely 3.26029.

Code for the Practice Exercises

1. `TableForm[Table[amts1[[i,2]],{i,1,3}]]`
2. ```
   TableForm[
      Table[
         {(amts1[[k,1]])^2, amts1[[k,2]]},
         {k,1,Length[amt1s]}
         ]
      ]
   ```

Problem

Below is data collected by Professor Clifton Albergotti, USF Physics Department, for an isotope of radioactive barium, barium-137. The first number in each pair is time in minutes. The second number is a Geiger counter reading of the emissions given off by a sample of this element.

Derive a formula in exponential form which fits the data and use it to determine the half-life of barium-137. Print out the derivation of the exponential formula and the estimation of half-life as well as a plot of the exponential function and the data points.

```
{{0, 2900}, {1, 2162}, {2, 1616}, {3, 1286},
 {4, 1018}, {5,  714}, {6,  635}, {8,  369},
 {9,  268}, {10,  210}, {11,  169}, {12, 121},
 {13,   93}, {14,  80}, {16,  48},  {17,  44},
 {18,  36}, {19,  21}
```

The data, entitled **rdecay.data** can also be found in the file folder **data** on your *Exploring Calculus* disk. It can be read into your notebook with a **ReadList** command.

`ReadList["rdecay.data", {Number, Number}]`
```

23

Projectile Motion in a Resisting Medium

In this exploration we're going to take another look at the problem of finding the trajectory of a projectile fired into the air. This same problem was investigated in Exploration Thirteen under the assumption that the motion took place in a vacuum. Here you will develop a more realistic model for projectile motion, one that takes into account the fact that the atmosphere exerts a force on the projectile which resists its motion.

The DSolve Command

Let's begin by looking at two warm-up problems which introduce the *Mathematica* command **DSolve**. Once we know how to use **DSolve**, we can put it to work finding projectile paths.

Example 1: Find a function $g(x)$ which passes through the point $(0,1)$ with a slope of 2 and which satisfies the differential equation $g''(x) = 5g'(x)$

This problem asks us to find an unknown function $g(x)$. We are given three clues to the nature of the function: its second derivative is five times its first derivative, it passes through the point $(0, 1)$, and it has a slope of 2 at that point. When we provide *Mathematica* with these clues via the **DSolve** command, it will return the function. (Warning: The symbol ' ' used to indicate the second derivative is a two key-stroke symbol. The prime key must be pressed twice. *The quotation mark will not work.*)

```
In[1]:=
        Clear[x,y,g]
        DSolve[
          {g''[x]==5g'[x], g'[0]==2, g[0]==1},
          g[x], x]
```

```
Out[1]=
                     3 + 2 E^5x
        {{g[x] -> ───────────── }}
                        5
```

Notice that **DSolve** takes three arguments. The first is a list {g''[x] == 5g'[x], g'[0] == 2, g[0] == 1} which contains the "clues". The second argument g[x] tells *Mathematica* that we want it to solve for $g(x)$. And the third argument x indicates that we want the rule for $g(x)$ to use x as the independent variable.

Check to make sure that we really got the function we were looking for.

```
In[2]:=
        g[x_]:= (3 + 2*E^(5*x))/5
```

```
In[3]:=
        g''[x]
```

```
Out[3]=
        10 E^5x
```

```
In[4]:=
        5g'[x]
```

```
Out[4]=
        10 E^5x
```

It looks good so far. The function's second derivative is five times its first. Now check the particular values for g and g'.

```
In[5]:=
        g[0]
```

```
Out[5]=
        1
```

```
In[6]:=
        g'[0]
```

```
Out[6]=
        2
```

Exactly right. The problem is solved.

Example 2: Find a function $h(x)$ which satisfies the differential equation $h''(x) + h(x) = 0$.

```
In[7]:=
        Clear[h,x]
        DSolve[h''[x] + h[x] == 0, h[x], x]
```

```
Out[7]=
        {{h[x] -> C[2] Cos[x] - C[1] Sin[x]}}
```

This **DSolve** output differs from the last in that the function returned contains the unspecified constants C[1] and C[2]. This means that you can formulate a solution to the differential equation by substituting any constant values for C[1] and C[2]. Here is an example:

```
In[8]:=
        h[x_]:= 3*Cos[x] - 5*Sin[x]
```

```
In[9]:=
        h''[x]
```

```
Out[9]=
        -3 Cos[x] + 5 Sin[x]
```

Check to make sure that the sum $h''(x) + h(x)$ is zero.

```
In[10]:=
        h''[x] + h[x]
```

```
Out[10]=
        0
```

Good. Another differential equation solved.

Practice Exercise 1: Find a solution to the differential equation $x^2 f''(x) + 2x f'(x) - 2f(x) = 0$ which passes through the point (0,2) with a slope of 5. (*Ans.* $f(x) = (3 + x^3*2 - 2)/(3*x^2)$)

Projectile Motion

Here is the problem we addressed in Exploration Thirteen.

> A projectile is fired at an angle of 30 degrees to the horizontal with an initial velocity of 1000 feet per second. Describe the path of the projectile.

In that exploration we found the parametric equations for the path of motion assuming that the only force acting on the projectile was gravity. This force affected only the vertical motion, so we knew that the horizontal velocity would not change. It had the same value throughout the flight of the projectile as it had initially. Such is not the case with the model we are studying here. Air resistance acts horizontally as well as vertically and so will slow both of these motions simultaneously.

Our goal here is to reformulate the parametric equations $x(t)$, $y(t)$ of the projectile's path to take into account the retarding effect of the medium through which the projectile is flying. As before, we will assume that the point of fire is the origin, and so $x(0) = 0$ and $y(0) = 0$.

The projectile's initial horizontal velocity is $x'(0)$ and can be calculated as follows:

```
In[11]:=
        degrees = Pi/180
        N[1000Cos[30 degrees]]

Out[11]=
        866.025
```

Thus, $x'(0) = 866.025$ ft/sec. Here is the initial vertical velocity:

```
In[12]:=
        N[1000Sin[30 degrees]]

Out[12]=
        500.
```

$y'(0) = 500$ ft/sec

FINDING y[t]

The projectile's initial vertical velocity, 500 ft/sec, is being slowed by gravity at the rate of 32 ft/sec for each second the projectile is aloft. The velocity is also being slowed by air resistance, so the formula for the acceleration in the vertical direction must look something like this:

$$y''(t) = -32 - \text{deceleration due to air resistance.}$$

The problem now is to describe the effect of the air mathematically in some reasonable way. For the purposes of this model, we will assume that the retarding effect of the air is proportional to the velocity; in other words, the faster the object moves, the greater the air resistance. Based on this assumption, the formula for $y''(t)$ becomes

$$y''(t) = -32 - k\, y'(t).$$

where the exact value of the constant k depends on the density of the air through which the projectile is moving.

We are now ready to formulate the **DSolve** command for $y(t)$.

```
In[13]:=
        Clear[y,t];
        DSolve[
          {y''[t] == -32 - k y'[t],
          y'[0] == 500,
          y[0] == 0
          },
          y[t], t
          ]
```

Out[13]=
{{y[t] ->

$$\frac{4 \ (8 + 125 \ k)}{k^2} - \frac{4 \ (8 + 125 \ k)}{E^{kt} \ k^2} - \frac{32 \ t}{k}\}\}$$

Mathematica must now be taught the formula for $y(t)$ contained in the output cell above. It would be a real nuisance to type it in by hand, and such a chore is not necessary. *Mathematica* can translate the output cell into input format using a command called **Copy Output from Above**. It is in the **Action** menu under **Prepare Input.** All you have to do is position the horizontal cursor under the output cell and click. Then pull down the **Action** submenu **Prepare Input** and release on **Copy Output from Above**. *Mathematica* will then place the cell below on your screen.

```
        {{y[t] ->
            (4*(8 + 125*k))/k^2 -
            (4*(8 + 125*k))/(E^(k*t)*k^2) - (32*t)/k
          }}
```

Edit this cell so that it becomes the function definition: Remove the braces and change the arrow to : =. Enter the definition.

```
In[14]:=
        y[t_]:=
            (4*(8 + 125*k))/k^2 -
            (4*(8 + 125*k))/(E^(k*t)*k^2) -
            (32*t)/k
```

In a command like this one where a long algebraic expression must be split across two or more lines, it is important that you put something at the end of the lines to indicate you have not completed the expression yet. The cell shown here has minus signs at the end of the second and third lines, so *Mathematica* knows that there is more to come. For example, if you put the minus sign at the beginning of line 3, then the expression on line 2 would be complete, and *Mathematica* would assign it to y[t]. The remaining two lines would be interpreted as a second command.

FINDING x[t]

The only acceleration in the *x* direction is that due to air resistance.

$$x''(t) = -k\, x'(t)$$

Here is the **DSolve** calculation of *x(t)*:

```
In[15]:=
        Clear[x,t];
        DSolve[
          {x''[t] == -k x'[t],
          x'[0]==866.025,
          x[0] ==0
          },
          x[t], t
          ]
```

```
Out[15]=
                    866.025     866.025
          {{x[t] -> ─────────── - ─────────── }}
                        k          E^kt   k
```

Teach *Mathematica* the formula for *x(t)*.

```
In[16]:=
        x[t_]:= 866.025/k - 866.025/(E^(k*t)*k)
```

THE PATH OF THE PROJECTILE

Now, let's set a value for *k*, say 0.1, and see what the plot looks like. We will have to experiment a bit to get the *t*-interval set up properly.

```
In[17]:=
        k = 0.1;
        x[t_]:= 866.025/k - 866.025/(E^(k*t)*k);
        y[t_]:=
            (4*(8 + 125*k))/k^2 -
            (4*(8 + 125*k))/(E^(k*t)*k^2) -
            (32*t)/k;
        ParametricPlot[{x[t],y[t]},{t,0,15}];
```

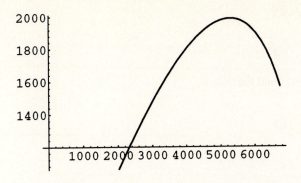

We need a **PlotRange** setting to force *Mathematica* to show us the origin.

```
In[18]:=
        ParametricPlot[
          {x[t],y[t]},  {t,0,15},
          PlotRange->{{0,8000},  {0,2000}}
          ];
```

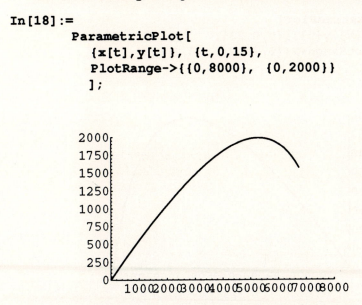

The *t*-interval isn't long enough to show the projectile returning to earth. Figure out how long the flight takes. It looks like it might be around 18 seconds.

```
In[19]:=
        Clear[t];
        t1 = FindRoot[y[t]==0,  {t,18}]
```

```
Out[19]=
        {t -> 23.0751}
```

Check it.

```
In[20]:=
        y[t]/.t1
```

```
Out[20]=
        -8.88178 10^{-16}
```

Close enough to zero. Find the range.

```
In[21]:=
        x[t]/.t1
```

```
Out[21]=
        7798.48
```

Redraw the plot to show the entire path.

```
In[22]:=
        ParametricPlot[
          {x[t],y[t]}, {t,0,23},
          PlotRange->{{0,7800},  {0,2000}}
          ];
```

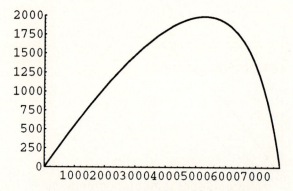

You can see the changes in the speed of the projectile by plotting its positions at equal time intervals.

```
In[23]:=
        Clear[positions];
        positions = Table[{x[t],y[t]},  {t,0,23}];
        ListPlot[
          positions,
          PlotStyle->{
            PointSize[0.03],
            Map[Point,positions]
            },
```

```
AxesLabel->{"x","y"},
PlotLabel->
   " Projectile Motion\n   k = 0.1",
PlotRange->{{0,7800},{0,2200}},
Ticks->
   {{0,7800},  {0,500,1000,1500,2000}}
];
```

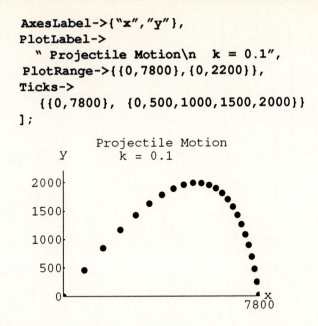

You can easily see the asymmetry of the path caused by the air resistance.

Code for the Practice Exercises

```
1. DSolve[
      x^2 f''[x] + 2x f'[x] - 2f[x]==0,
      f[x], x
      ]
```

Problems

1. Solve the differential equation $y''(t) = -y'(t)$, $y'(0) = 3$, and $y(0) = 2$ for $y(t)$.

2. Find the path of a projectile fired at an angle of 45 degrees to the horizontal with an initial velocity of 1000 feet per second. How does the range of this projectile compare with that worked out for a 30 degree angle of fire? Use $k = 0.1$.

3. Use **ListPlot** to graph paths for the 30 degree angle of fire and initial velocity 1000 feet per second. Use $k = 10, 1, 0.1, 0.01$, and 0.001. Write a paragraph explaining how changes in k alter the shape of the path.

4. Using $k = 0.1$, and the initial velocity 1000 feet per second, experiment with the angle to find the one that gives the greatest horizontal distance. Is the angle larger or smaller than the angle that gives the greatest distance when air resistance is ignored?

24

Surfaces

In this section you will learn how to use *Mathematica* graphics commands to represent surfaces in three dimensions.

The Plot3D Command

The **Plot3D** command graphs surfaces of the form $z = f(x, y)$. The simplest version of the command is illustrated below.

```
In[1]:=
        Plot3D[5 - x^2 - y^2, {x,-3,3}, {y,-4,4}];
```

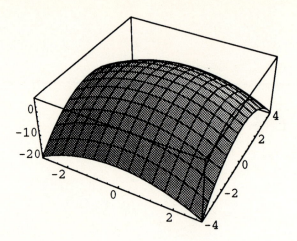

PLOTRANGE

The lists {x, −3, 3} and {y, −4, 4} determine a rectangular region in the *xy*-plane over which the surface will be drawn. The vertical scale is set automatically, but it can also be controlled using a **PlotRange** setting.

```
In[2]:=
        Plot3D[
            5 - x^2 - y^2,
            {x,-3,3},  {y,-4,4},
            PlotRange->{-20,20}
            ];
```

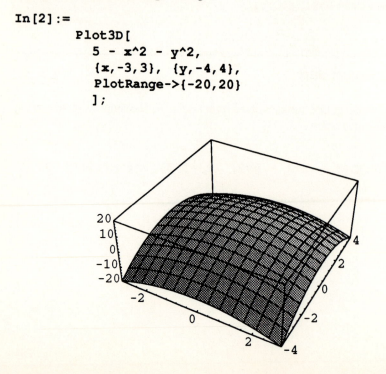

BOXRATIOS AND VIEWPOINT

A box whose default ratios are 1:1:0.4 is drawn around the surface, and you appear to be looking at the surface from above the box. In fact, if you think of the box as having sides of length 1, then the viewpoint is (1.3, –2.4, 2). The box ratios and the viewpoint can both be changed. Try this adjustment to the plot above:

```
In[3]:=
        Plot3D[
            5 - x^2 - y^2,
            {x,-3,3}, {y,-4,4},
            PlotRange->{-20,20},
            BoxRatios->{1,1,1}
        ];
```

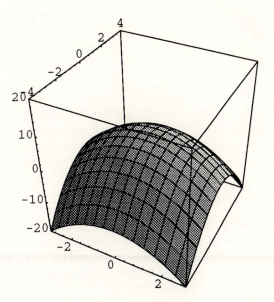

Here is the plot with a different viewpoint:

```
In[4]:=
        Plot3D[5 - x^2 - y^2, {x,-3,3}, {y,-4,4},
            PlotRange->{-20,20},
            BoxRatios->{1,1,1},
            ViewPoint->{0,2,0}];
```

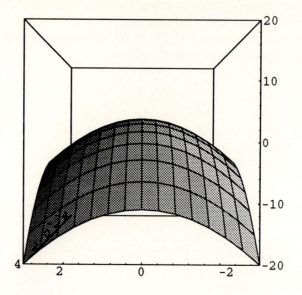

Clearly, the **ViewPoint** command makes an enormous difference in the way the surface looks. However, it is not at all clear how to choose the values for this option. Fortunately, the Macintosh interface for *Mathematica* provides a visual method for selecting the viewpoint.

THE 3D VIEWPOINT SELECTOR

In order to use the selector, write out your plot command *without* setting the **ViewPoint** option. Leave the cursor at the point where you want to insert this option. Hold the ⌘-key and the shift key down simultaneously and press the v-key. A dialog box will appear. Alternatively, the dialog box can be found in the **Action** menu under **Prepare Input**.

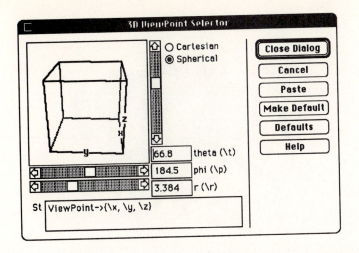

The dialog box contains a cube whose position can be changed using the three scroll bars. Move them to orient the cube as you wish. When it is in position, click on **Paste**. The **ViewPoint** option will be set in your **Plot3D** command at the place where you left the cursor.

If you click **Make Default** in this dialog box, then the current viewpoint will be used in all subsequent plots unless you specify another value. To return to the original viewpoint click the **Defaults** button. You may wish to select Cartesian coordinates instead of Spherical. It doesn't really matter which you use, because either way you can orient the cube the way you want it.

Practice Exercise 1: Use the **ViewPoint Selector** to adjust your plot so that you see the surface from below; directly from the side.

Practice Exercise 2: Redraw the plot with the settings **Boxed**→**False** and **Axes**→**None** to see the effect of these options.

Let's try another plot, $z = \sin(x^2 + y^2)$. Label the plot and the axes.

```
In[5]:=
      Plot3D[
         Sin[x^2 + y^2],
         {x,-3,3},  {y,-3,3},
         PlotLabel->"z = sin(x^2+y^2)",
         AxesLabel->{"x","y"," "}
         ];
```

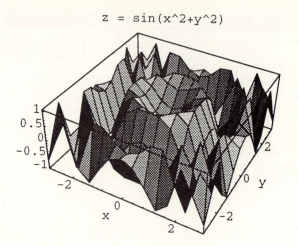

PLOTPOINTS

It is hard to believe that a function as smooth as the sine would produce such a craggy plot. The problem is this: Surfaces are drawn by taking 15 x-values and 15 y-values and using them to find 225 function values from which the surface is rendered. It looks like the 15 by 15 grid is too coarse for this plot. With the **PlotPoints** option you can choose a finer grid size. However, be aware that setting this option to a large number will cause the plotting to go very slowly and will use up lots of memory. It would be a good idea to **Clear** any previous 3D plots from your *Mathematica* notebook before entering the next command.

```
In[6]:=
    Plot3D[
       Sin[x^2 + y^2],
       {x,-3,3}, {y,-3,3},
       PlotLabel->"z = sin(x^2+y^2)",
       AxesLabel->{"x","y"," "},
       PlotPoints->40
       ];
```

z = sin(x^2+y^2)

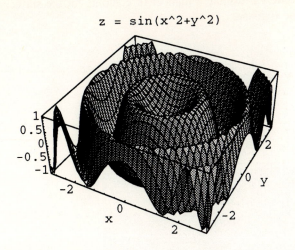

That looks better.

Contour Plots

A second way to visualize a surface is to draw a contour plot (or contour map). A contour is the curve in which a plane parallel to the xy-plane (z = constant) cuts through the surface. In the picture below you can see that the plane $z = -3$ intersects the surface $z = 5 - x^2 - y^2$ in a circle. The circle is referred to as the surface contour for $z = -3$.

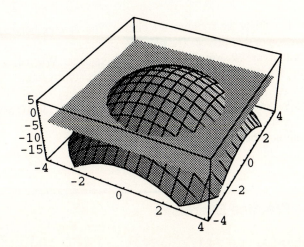

The **ContourPlot** command will display a number of contours for the surface, drawing them all together in the *xy*-plane.

```
In[7]:=
        ContourPlot[
          5 - x^2 - y^2,
          {x, -4,4}, {y,-4,4}
          ];
```

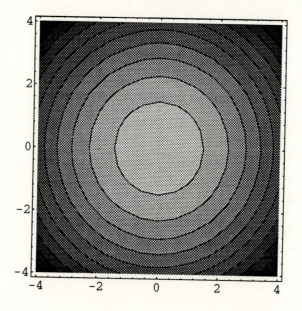

You can see that all of the contours are circles. The shading between the circles indicates the elevation of the surface. The paler the shading, the higher the point on the surface. When contours are close together, the surface is steeper. When they are far apart, the surface is more level. For this plot the highest point is at the center.

Practice Exercise 3: Draw a contour plot for the sine surface we looked at earlier.

Clear all your plots before going on to the problems.

Code for the Practice Exercises

1. `ViewPoint->{0, 0, -2},`
   `ViewPoint->{-3.6, 0.087, 0.524}`

3. The following command does a pretty good job, but it takes a while.

```
ContourPlot[
   Sin[x^2 + y^2],
   {x,-3,3}, {y,-3,3},
   PlotPoints->25
   ];
```

Problems

1. Make a three-dimensional plot of each of the following surfaces. Choose appropriate options so that the significant aspects of the surface are clear. Make your adjustments on the screen; only print the plot when you are satisfied with it. Include a label with each surface. **Clear** each plot before going on to the next one.

   (a) $z = x^2 + y^2$
   (b) $z = x^2 - y^2$
   (c) $z = \sin(x^2 + y^2)^{1/2}/(x^2 + y^2)^{1/2}$
   (d) $z = x\, e^{(x2 - y2)}$
   (e) $z = \cos((x^2 + y^2)^{1/2})$

2. Make contour plots for each of the surfaces in Problem 1.

25

3D Critical Points

In this section you will use *Mathematica* to find the critical points on a surface; to characterize the points as local maximum, minimum, or saddle points; and to draw a graph of the surface which displays them.

For example, enter the function $z = x^4 - 8xy + 2y^2 - 3$ and calculate its first partial derivatives.

In[1]:=
```
Clear[x,y,z];
z = x^4 - 8x y + 2y^2 - 3
```

Out[1]=
$$-3 + x^4 - 8 x y + 2 y^2$$

In[2]:=
```
D[z,x]
```

Out[2]=
$$4 x^3 - 8 y$$

In[3]:=
```
D[z,y]
```

Out[3]=
$$-8 x + 4 y$$

To find the critical points all we have to do is set both these derivatives equal to 0 and solve for x and y. Call the list of solutions critpts so we can refer back to it easily.

```
In[4]:=
        critpts = Solve[
          {D[z,x]==0,
          D[z,y]==0},
          {x,y}
          ]
```

```
Out[4]=
        {{x -> 2, y -> 4}, {x -> -2, y -> -4}, {x -> 0, y
        -> 0}}
```

Find the value of z at each of these critical points.

```
In[5]:=
        z/.critpts
```

```
Out[5]=
        {-19, -19, -3}
```

So, now we have three critical points $(2, 4, -19)$, $(-2, -4, -19)$, and $(0, 0, -3)$. A critical point that is isolated (that is, one that is not part of a curve of critical points) will be a local maximum, a local minimum, or a saddle point. We'd like to figure out what sort of points we have here for z. One way to find out is to graph the surface near the point and see what it looks like there. It may take several attempts to get the surface oriented in such a way that its behavior is clear. Here is a first stab at a plot.

```
In[6]:=
        Plot3D[z, {x,-3,3}, {y,-5,5}];
```

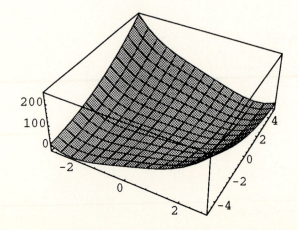

It is hard to believe there is anything much going on there. It's a good thing we already know about the critical points. We just have to force the plot to show them. The problem is probably in the PlotRange setting. It runs from –20 or so all the way up to 200, so the scale can't register the difference between a z-value of –19 and one of 0 very clearly. Try this.

```
In[7]:=
        Plot3D[
          z, {x,-3,3}, {y,-5,5},
          PlotRange->{-20,0}
          ];
```

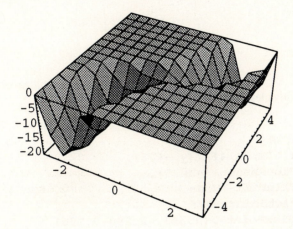

Mathematica sliced off the surface at $z = 0$ and filled in the top of the box. Get rid of that fill.

```
In[8]:=
        Plot3D[
          z, {x,-3,3}, {y,-5,5},
          PlotRange->{-20,0},
          ClipFill->None
          ];
```

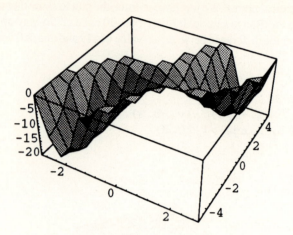

With a little experimentation, we came up with option settings which show the critical points pretty clearly.

```
In[9]:=
        Plot3D[
          z, {x,-4,4}, {y,-6,6},
          PlotRange->{-20,0},
          ClipFill->None,
          PlotPoints->25,
          ViewPoint->{2.363,-2.290,0.790}
          ];
```

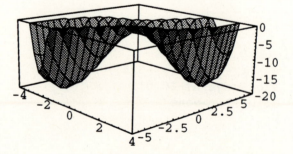

The points $(2, 4, -19$ and $(-2, -4, -19)$ are local minima, and the point $(0, 0, -3)$ is a saddle point.

A contour plot can also be useful in deciding on a function's behavior at critical points. But, once again, some experimentation is usually necessary in order to get the best picture. This plot does not tell us much.

```
In[10]:=
        ContourPlot[z,  {x,-4,4},  {y,-6,6}];
```

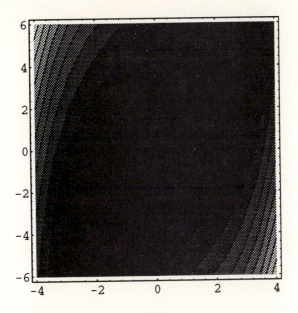

This next contour plot is much more revealing.

```
In[11]:=
        ContourPlot[z,  {x,-3,3},  {y,-6,6},
            PlotPoints->20,
            PlotRange->{-20,0}];
```

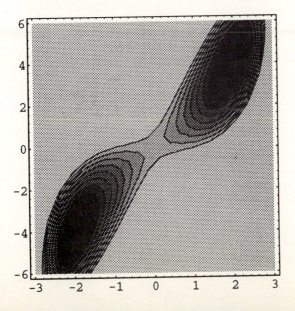

You can see that the points $(-2, -4)$ and $(2, 4)$ are surrounded by contour lines. That means that the points are optimum points (maximum or minimum points, not saddle points). The very dark shading tells us that they are both minima. There are no contour lines surrounding $(0, 0)$, so it is a saddle point.

Problems

For each of the following functions, find the critical points and identify each as a local maximum, local minimum, or a saddle point. Print out your best three-dimensional plot and contour plot for each.

1. $z = 4\,x*y - x^4 - y^4$

2. $z = x^3 + y^3 + 3\,x^2 - 3y^2 - 8$

3. $z = x *\sin(y)$

   Comment: The Solve command won't be much help. Try making plots to see what the function is doing.

4. $z = 2\,x*y \,/(\,x^2 + y^2)$

   Comment: This is a trick question. See if you can see why.

26

Constrained Optimization in Two Variables

For a function of a single variable finding an absolute maximum or minimum on a closed interval usually involves (1) finding the critical points in the interval, and (2) evaluating the function at each critical point *and* at the endpoints of the interval. Then comparing these function values determines the function's optimum values. For a function of two variables the procedure for finding the absolute maximum or minimum on a restricted region is similar: (1) Find the critical points inside the region of the xy-plane over which the function is being studied, and (2) look for possible maximum or minimum values on the boundary of the region. The first step was discussed in a previous exploration. It usually requires solving a system of two equations in two unknowns, which come from setting the partial derivatives to 0. The second step can take more effort than you would expect. You must consider each part of the boundary; find any maximum or minimum values on it; check each point from the boundary against the critical points from inside the region; and finally choose the largest or the smallest function value generated by these points.

We start by considering the problem of finding optimum values for a function of two variables when the (x, y) points are required to be on a curve in the xy-plane. This is, we want to find the optimum points on the boundary of a region. Then we will consider the problem of finding optimum values on an entire region in the xy-plane.

Optimization on a Curve

Consider a specific problem:

> Find the absolute maximum and minimum values taken by $f(x, y) = x^2 - y^2$, where the points (x, y) are required to be on the circle $x^2 + y^2 = 4$.

Begin by defining the function.

```
In[1]:=
        Clear[f,x,y];
        f[x_,y_]:= x^2 - y^2
```

Take a look at the surface.

```
In[2]:=
        Plo   3D[
          f[x,y],{x,-3,3},{y,-3,3},
          AxesLabel->{"x","y"," "},
          Plo   Label->"z = x^2 - y^2",
          BoxRa  ios->{1,1,1}
          ];
```

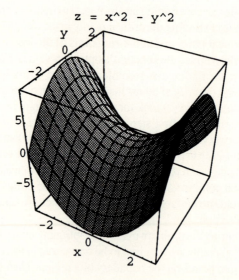

There are no global optimum points on this surface. It has only one critical point, the origin, and that is a saddle point. However, in this problem we are not concerned with every point on the surface. We are only interested in those surface points whose x and y coordinates satisfy the equation $x^2 + y^2 = 4$. The plot of this equation in the xy-plane is a circle centered at the origin with a radius of 2. So, the surface points we are interested in are those which lie directly above or below the circle. Here is a sketch of them displayed with the surface.

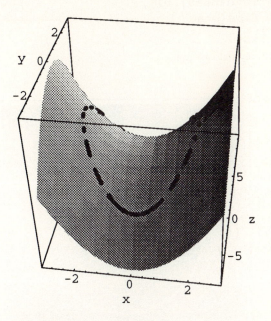

We want to find the optimum points on this curve.

THE CONTOUR PLOT METHOD.

It is a little hard to see where the circle $x^2 + y^2 = 4$ is in relation to this picture because the surface obscures most of the the xy-plane. One way to view the surface and the circle together is to render the surface in two dimensions using **ContourPlot** and then draw the circle on top of it. This can be done in *Mathematica* although the procedure is a bit awkward. We must first generate a table of points for the circle and then use an **Epilog** command to to join up the points and place the resulting curve in the contour plot. The code is given below.

```
In[3]:=
    pts = Table[
        N[{2 Cos[t], 2 Sin[t]}],
        {t,0,2 Pi, Pi/10}
        ];
```

```
ContourPlot[
  f[x,y],{x,-3,3},{y,-3,3},
  Epilog->{
    GrayLevel[0.9],
    Thickness[.01],
    Line[pts]
    }
  ];
```

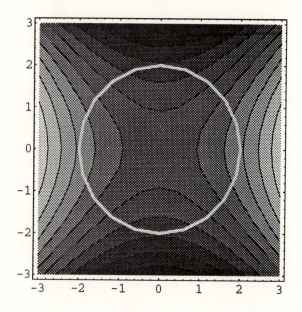

Recall that in a contour plot the palest points correspond to the the highest function values, the darkest points the lowest. With this in mind it is fairly clear that the maximum and minimum values are taken at the four points $(0, 2), (2, 0), (-2, 0),$ and $(0, -2)$. Calculate the function values at these points.

```
In[4]:=
    f[0,2]
```

```
Out[4]=
    -4
```

```
In[5]:=
    f[2,0]
```

```
Out[5]=
    4
```

```
In[6]:=
        f[0,-2]
```

```
Out[6]=
        -4
```

```
In[7]:=
        f[-2,0]
```

```
Out[7]=
        4
```

So, the absolute maximum is 4 at (2, 0) and (–2, 0), and the absolute minimum is –4 at (0, 2) and (0, –2).

Practice Exercise 1: Estimate the maximum and minimum values taken on by the function $f(x, y) = x^2 - y^2$ on the line segment $y = -2x + 6$, for $-2 \le x \le 8$ by making a contour plot and drawing the line on it. Note that you only need the two endpoints to draw the line segment. (*Ans.* 12 when $x = 4$, and –96 when $x = -2$)

REDUCTION TO ONE VARIABLE

Here is an approach to this type of optimization that turns it into a one-variable calculus problem. The curve $x^2 + y^2 = 4$ can be written parametrically as

$$x = 2\cos(t), \quad y = 2\sin(t),$$

for $0 \le t \le 2\pi$. Substitute these values into $z = f(x, y)$, and we get z as a function of t.

```
In[8]:=
        Clear[z,t]
        z[t_]:= f[2 Cos[t],2 Sin[t]]
```

As t increases in value from 0 to 2π the point (x, y) on the circle traces a complete circumference, and the corresponding point on the surface moves through all of its possible positions. If you plot z against t, you'll see the maximum and minimum z-values quite clearly.

```
In[9]:=
        Plot[
          z[t], {t,0,2 Pi},
          AxesLabel->{"t","z"}
          ];
```

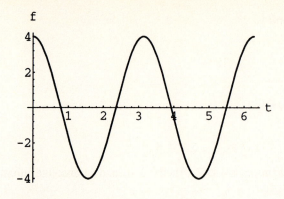

Find the values of t at which z takes its maximum and minimum values by setting the derivative of z equal to zero. From the plot you can see that they are near 0, 1.5, 3, and 4.5:

```
In[10]:=
        FindRoot[z'[t]==0,{t,0}]

FindRoot::frmp:
            Machine precision is insufficient to
                achieve the accuracy 1. 10⁻⁹.
```

$$\text{FindRoot::frmp:}$$

```
Out[10]=
        {t -> 0.}
```

The corresponding surface point is $(2 \cos(t),\ 2 \sin(t),\ z(t))$. The error message indicates that the result is not as accurate as the default settings in *Mathematica* request; however the accuracy is sufficient for our purposes.

```
In[11]:=
        {2 Cos[t], 2 Sin[t], z[t]}/.%

Out[11]=
        {2., 0., 4.}
```

Repeat the same steps for the other three points.

```
In[12]:=
        FindRoot[z'[t]==0,{t,1.5}]

Out[12]=
        {t -> 1.5708}

In[13]:=
        {2 Cos[t], 2 Sin[t], z[t]}/.%

Out[13]=
        {-5.42101 10⁻²⁰, 2., -4.}
```

```
In[14]:=
        FindRoot[z'[t]==0,{t,3}]
```

```
Out[14]=
        {t -> 3.14159}
```

```
In[15]:=
        {2 Cos[t], 2 Sin[t], z[t]}/.%
```

```
Out[15]=
        {-2., -1.0842 10⁻¹⁹, 4.}
```

```
In[16]:=
        FindRoot[z'[t]==0,{t,4.5}]
```

```
Out[16]=
        {t -> 4.71239}
```

```
In[17]:=
        {2 Cos[t], 2 Sin[t], z[t]}/.%
```

```
Out[17]=
        {3.79471 10⁻¹⁹, -2., -4.}
```

These are essentially the same points we found earlier with the **ContourPlot** technique; the numbers with 10^{-19} are virtually zero.

Practice Exercise 2: Find the maximum and minimum values taken by $f(x, y) = x^2 - y^2$ on the line segment $y = -2x + 6$, for $-2 \leq x \leq 8$ by writing z in terms of x only, and then using the one variable techniques. (*Ans.* 12, -96)

Optimization on a Region

Now consider the problem of finding maximum and minimum values for a function of two variables when the variables are restricted to a region of the xy-plane. Consider a specific example.

> Find the absolute maximum and minimum of
>
> $$g(x, y) = x^2 + y^2 - 4x - 2y + 7$$
>
> on the semicircular region bounded by the x axis and $x^2 + y^2 = 16$ in the upper half-plane.

Look at a contour map together with the boundary curves for the region. The semicircle can be parametrized by $x = 4 \cos(t)$, $y = 4 \sin(t)$, for $0 \leq t \leq \pi$. First define the function.

```
In[18]:=
       Clear[g,x,y];
       g[x_,y_]:= x^2 + y^2 - 4 x - 2 y + 7
```

Then make a table of values along the boundary, to use for plotting.

```
In[19]:=
       pts = Table[
          N[{4 Cos[t], 4 Sin[t]}],
          {t,0,Pi,Pi/10}
          ];
```

```
In[20]:=
       ContourPlot[
          g[x,y], {x,-5,5}, {y,-1,5},
          Epilog->{
             GrayLevel[0.8],
             Thickness[0.01],
             Line[pts],
             Line[{{-4,0},{4,0}}]
             }
          ];
```

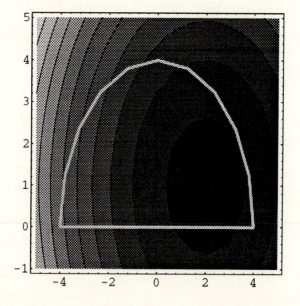

It looks like there is a critical point inside the region, at the center of the elliptical contour; and another extreme value is at the corner of the region at (–4, 0). Use partial derivatives to find the critical point.

```
In[21]:=
        Solve[
            {D[g[x,y],x]==0,  D[g[x,y],y]==0},
            {x,y}
            ]
```

```
Out[21]=
            {{x -> 2,  y -> 1}}
```

```
In[22]:=
        g[x,y]/.%
```

```
Out[22]=
            {2}
```

Checking the values along the boundary is a problem like the ones in the first part of this exploration; we must find the function's optimum values on the curves which bound the region. When this task is complete, the largest and smallest border values are compared to the values of the function at the interior critical point. As it turns out, in this particular problem, there is little work to do because it is obvious from the contour plot that the boundary's maximum function value occurs at $(-4, 0)$ and the minimum at $(2, 0)$. But in future problems the contour plot might not be so easy to read, so let's use calculus to verify our work.

Along the line at the bottom of the region, $y = 0$ and $-4 \le x \le 4$. The relationship between g and x along this part of the boundary can be visualized with the plot.

```
In[23]:=
        Plot[
            g[x,0],  {x,-4,4},
            PlotRange->All,
            AxesLabel->{"x","g"}
            ];
```

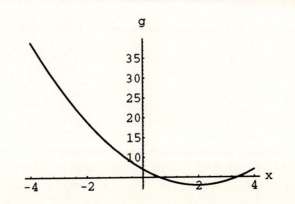

The maximum value occurs when $x = -4$; there is also a minimum near $x = 2$. To find it exactly, take the derivative of $g(x, 0)$ and set it to zero.

```
In[24]:=
        Solve[D[g[x,0],x]==0,  x]
```

```
Out[24]=
        {{x -> 2}}
```

The value of g at $x = 2$, $y = 0$ is:

```
In[25]:=
        g[2,0]
```

```
Out[25]=
        3
```

And at $x = -4$, $y = 0$ the value is

```
In[26]:=
        g[-4,0]
```

```
Out[26]=
        39
```

Now look at the function on the semicircle $x = 4\cos(t)$, $y = 4\sin(t)$, $0 \le t \le \pi$.

```
In[27]:=
        Plot[g[4 Cos[t],4 Sin[t]], {t,0,Pi}];
```

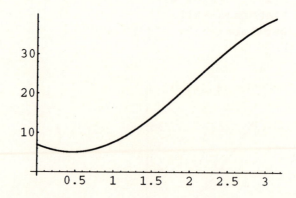

There is a maximum at $t = \pi$, which corresponds to $x = 4\cos(\pi) = -4$, $y = 4\sin(\pi) = 0$. And there is a minimum around $t = 0.5$. To find the exact value, set the derivative to zero and solve; since this is not a polynomial equation, use **FindRoot** instead of **Solve**.

As a function of t, g is

```
In[28]:=
        g[4 Cos[t],4 Sin[t]]
```

```
Out[28]=
        7 - 16 Cos[t] + 16 Cos[t]² - 8 Sin[t] +
           16 Sin[t]²
```

The derivative of g with respect to t is

```
In[29]:=
        D[%,t]
```

```
Out[29]=
        -8 Cos[t] + 16 Sin[t]
```

Set the derivative to 0 and solve for t.

```
In[30]:=
        FindRoot[-8Cos[t] + 16Sin[t]==0, {t,0.5}]
```

```
Out[30]=
        {t -> 0.463648}
```

The point is

```
In[31]:=
        {4 Cos[t], 4 Sin[t]}/.%
```

```
Out[31]=
        {3.57771, 1.78885}
```

And the value is

```
In[32]:=
        g[3.57771, 1.78885]
```

```
Out[32]=
        5.11145
```

Collecting together all of the points, we have

$x = 2$	$y = 1$	$g = 2$	<——Minimum
$x = 2$	$y = 0$	$g = 3$	
$x = -4$	$y = 0$	$g = 39$	<——Maximum
$x = 3.57771$	$y = 1.78885$	$g = 5.11145$	

Code for the Practice Exercises

```
1. ContourPlot[
     x^2 - y^2, {x,-3,9}, {y,-12,0},
     Epilog->{
       GrayLevel[0.5],
       Thickness[0.01],
       Line[{ {-2,2}, {8,-12} }]
       }
     ];

   {x,-2x+6,z[x,-2x+6]}/.{x->-2}
   {x,-2x+6,z[x,-2x+6]}/.{x->8}
2. Clear[x,y,z];
   z[x_,y_]:= x^2 - y^2;
   Solve[D[z[x,-2x+6],x]==0,x]
   {x, -2x+6, z[x,-2x+6]}/.%
   {x, -2x+6, z[x,-2x+6]}/.{x->-2}
   {x, -2x+6, z[x,-2x+6]}/.{x->8}
```

Problems

1. Find the absolute maximum and minimum values of $f(x, y) = 2x^2 + y^2$ on the curve
   $y = 10 - 5x$ for $0 \leq x \leq 2$. Make a contour plot with the curve on it.

2. Find the absolute maximum and minimum values of $f(x, y) = 2x + 3y$ on the semicircle
   $x = 5\cos(t)$, $y = 5\sin(t)$, for $0 \leq t \leq \pi$. Make a contour plot with the curve on it.

3. Find the absolute maximum and minimum values of $f(x, y) = x^2 + xy + y^2 - 6x$ on the
   rectangular region $0 \leq x \leq 5$, $-3 \leq y \leq 3$. Make a contour plot with the curve on it.

4. Find the absolute maximum and minimum values of $f(x, y) = x^2 - xy + y^2 + 1$ on the
   triangular region in the first quadrant bounded by the lines $x = 0$, $y = 4$, and $y = x$. Make a
   contour plot with the curve on it.

Appendix 1

Glossary of *Mathematica* Commands

This appendix contains a list of the *Mathematica* commands introduced in this text together with brief descriptions of their function and format. These descriptions are restricted to the uses made of the commands in the various Explorations. This means that in some cases certain features of a command may not be completely described. Characteristics or options not appropriate for beginning calculus have been omitted. Full descriptions of these commands and the hundreds of other functions available to you in *Mathematica* can be found in the manual, *Mathematica, A System for Doing Mathematica by Computer*, Stephen Wolfram, Addison-Wesley Publishing Company, Inc.

AspectRatio
AspectRatio is an option for two-dimensional graphics commands which specifies the ratio of height to width for a plot. For example, **AspectRatio–>Automatic** used in a **Plot** command will produce a graph in which the scales on the horizontal and vertical axes are the same. Refer to Exploration Thirteen.

Automatic
When a graphics option is set to **Automatic**, *Mathematica* will use its internal algorithms to determine a setting for the option. Refer to Exploration Eight.

AxesLabel
AxesLabel is an option that can be set in a graphics command to specify labels for the axes. **AxesLabel–>{***xlabel*, *ylabel***}** labels the horizontal and vertical axes for a two-dimensional plot. **AxesLabel–>{***xlabel*, *ylabel*, *zlabel***}** labels a three-dimensional plot.

AxesLabel–>None specifies that no labels should be given. **AxesLabel–>***label*, labels the vertical axes of a two-dimensional plot and the third axis of a three-dimensional plot. Refer to Explorations Two, Twenty-four.

AxesOrigin

AxesOrigin is an option for two-dimensional graphics functions which specifies the point where the horizontal and vertical axes cross. For example, **AxesOrigin–>{–3, 5}**, will cause the axes to cross at the point (–3, 5). Refer to Exploration Five.

Block

When the command **Block[{x, y, ...}, *expression*]** is entered, *Mathematica* will execute the code contained in *expression*. First, however, any values previously assigned to the variables *x, y*,... are automatically cleared. Then, when the execution of the **Block** is finished, the original values of these symbols are restored. **Block** is automatically used to localize values of the iterators in **Do**, **Sum**, and **Table** commands. Refer to Explorations Seventeen, Twenty.

BoxRatios

BoxRatios is an option for three-dimensional graphics which specifies the relative length of the edges of the bounding box of the three-dimensional plot. Refer to Exploration Twenty-four.

Clear

Clear[*symbol$_1$ symbol$_2$* ...]** clears all values and definitions previously assigned to the symbols listed. Refer to Exploration Two.

ClipFill

ClipFill is an option for three-dimensional plots which specifies what is to be shown in cases where the surface would extend beyond the bounding box. For example, **ClipFill–>None** make holes in the surface where the bounding box intersects it. Refer to Exploration Twenty-five.

ContourPlot

ContourPlot[*f*, {*x, xmin, xmax*}, {*y, ymin, ymax*}]** generates a contour plot of *f* as a function of *x* and *y* over the region specified by the ranges on *x* and *y*. Refer to Exploration Twenty-four.

Cos

Cos[*z***]** gives the cosine of *z* radians. Refer to Exploration Twelve.

D

D[*f, x***]** gives the derivative or partial derivative of *f* with respect to *x*. **D[***f*, {*x, n*}]** gives the *n*th derivative or partial derivative of *f* with respect to *x*. Refer to Exploration Twenty-five.

Dashing

Dashing[{*r$_1$, r$_2$*, ... }]** is a two-dimensional graphics setting which specifies that lines are to be drawn dashed, with successive segments of length $r_1, r_2, ...$ (repeated cyclically). Refer to Exploration Seven.

Denominator
Denominator[*expression*] returns the denominator of *expression*. Refer to Exploration Seven.

Direction–>1
When used in a limit command, **Limit[f[x], x–>a, Direction–>1]**, causes x to approach a in the direction of the vector i, in other words, from the left.

Direction–>–1
When used in a limit command, **Limit[f[x], x–>a, Direction–>–1]**, causes x to approach a in the direction of the vector $–i$, in other words, from the right.

DisplayFunction–>Identity
This setting for **DisplayFunction** used in a graphics command suppresses the display of the plot. Refer to Exploration Eight.
DisplayFunction–>$DisplayFunction
This setting for **DisplayFunction** used in a graphics command causes the plot to be displayed. Refer to Exploration Eight.

Do
Do[*expression*, {*i*, *imin*, *imax*, *di*}] evaluates *expression* beginning with $i = imin$ and ending with $i = imax$ using increments of *di*. Refer to Exploration Eleven.

DSolve
DSolve[{*equation*$_1$, *equation*$_2$, ...}, *y*[*x*], *x*] solves a list of differential equation for the function $y(x)$ with independent variable x. Refer to Exploration Twenty-three.

E
Symbol for the number $e = 2.71828...$. Refer to Exploration Twenty-one.

Epilog
Epilog is an option for graphics functions which gives a list of graphics primitives to be drawn in after the main part of the graphics is rendered. Refer to Exploration Nine.

Expand
Expand[*expression*] multiplies out the products and positive integer powers in *expression*. Refer to Exploration One.

Factor
Factor[*polynomial*] factors a polynomial. It only works on exact integer or rational coefficients. Refer to Exploration One.

FactorInteger
FactorInteger[*n*] gives a list of prime factors of the integer n, together with their exponents. Refer to Exploration One.

FindRoot
FindRoot[*leftside* == *rightside*, {*x*, *x*$_0$}] solves the equation *leftside* == *rightside* numerically for x beginning with x_0. Refer to Exploration Twenty-one.

Fit

Fit[*data*, {*x*, 1}, *x*] fits the points in *data* with a least squares line. It returns the equation of the line using *x* as the independent variable. Refer to Exploration Twenty-two.

GrayLevel

GrayLevel[*n*], where $0 \leq n \leq 1$, sets the intensity of the gray for a black and white plot. Setting *n* to 0 produces a white plot, to 1 a black plot. Refer to Explorations Five, Nine.

Infinity

Infinity is a symbol for positive infinity. Refer to Exploration Seven.

Integrate

Integrate[*f*, {*x*, *xmin*, *xmax*}] returns a value for the definite integral of *f* for *x* between *xmin* and *xmax*. **Integrate**[*f*, *x*] returns the antiderivative of *f* with respect to *x*. Refer to Explorations Fifteen, Seventeen.

Length

Length[*expression*], gives the number of elements in *expression*. Refer to Exploration Twelve.

Limit

Limit[*expression*, *x*–>*x*$_0$], returns the limit of *expression* as *x* approaches *x*$_0$. Refer to Exploration Three.

Line

Line[{{*x*$_0$,*y*$_0$}, {*x*$_1$,*y*$_1$}, ..., {*x*$_n$,*y*$_n$}}], is a graphics primitive which represents line segments joining the sequence of points {*x*$_0$,*y*$_0$}, {*x*$_1$,*y*$_1$}, ..., {*x*$_n$,*y*$_n$} according to their order in the list. The command **Show**[**Line**[{{*x*$_0$,*y*$_0$}, {*x*$_1$,*y*$_1$}, ... ,{*x*$_n$,*y*$_n$}}]], for example, causes the line segments to be drawn. Refer to Explorations Fourteen and Seventeen.

ListPlot

ListPlot[{{*x*$_0$,*y*$_0$},{*x*$_1$,*y*$_1$},...{*x*$_n$,*y*$_n$}}] graphs the points in the list {{*x*$_0$,*y*$_0$},{*x*$_1$,*y*$_1$},...{*x*$_n$,*y*$_n$}}. Options available include **AspectRatio**, **AxesLabel**, **DisplayFunction**, **PlotLabel**, **PlotRange**, **PlotStyle**, **Ticks**. Refer to Exploration Two.

Log

Log[*n*] returns the natural logarithm of *n*. **Log**[*b*, *n*] returns the log of *n* to the base *b*. Refer to Exploration Twenty-two.

Map

Map[*f*, *list*] applies the function *f* to each element in *list*. Refer to Exploration Seventeen.

N

N[*expression*] gives the numerical value of *expression* in decimal form. **N**[*expression*, *n*] does the computations to *n*-digit precision. Refer to Exploration One.

NIntegrate

NIntegrate[*f*, {*x*, *xmin*, *xmax*}] returns a numerical value for the integral of *f* for *x* between *xmin* and *xmax*. Refer to Exploration Seventeen.

Numerator
Numerator[*expression*] gives the numerator of *expression*. Refer to Exploration Seven.

ParametricPlot
ParametricPlot[{*x(t)*, *y(t)*}, {*t*, *tmax*, *tmin*}] draws the parametic curve $x = x(t)$, $y = y(t)$ for the indicated range on *t*. The options which can be set for **ParametricPlot** are the same as those for **Plot**. Refer to Exploration Three.

%
Symbol for the previous output. **%**n specifies output number n. Refer to Exploration One.

Pi
The symbol for π. Refer to Exploration One.

Plot
Plot[*f(x)*, {*x*, *xmin*, *xmax*}] graphs *f* as a function of *x* for the indicated range. The following options for **Plot** are discussed in this book: **AspectRatio**, **AxesLabel**, **AxesOrigin**, **DisplayFunction**, **Epilog**, **PlotLabel**, **PlotRange**, **Prolog**, **Ticks**. For a complete list of options see the *Mathematica* manual. Refer to Exploration One.

Plot3D
Plot3D[*f(x, y)*, {*x*, *xmin*, *xmax*}, {*y*, *ymin*, *ymax*}] graphs *f* as a function of *x* and *y* over the indicated region. The following options for **Plot3D** are discussed in this book: **AspectRatio**, **AxesLabel**, **Boxed**, **BoxRatios**, **Epilog**, **PlotLabel**, **PlotRange**, **PlotPoints**, **Prolog**, **Ticks**, **ViewPoint**. For a complete list of options see the *Mathematica* manual. Refer to Exploration Twenty-four.

PlotLabel
PlotLabel–>*label* specifies a label for the plot. Such a label can be given in addition to or instead of labels for the coordinate axes. Refer to Exploration Nine.

PlotPoints
PlotPoints–>*n* specifies the number of sample points to use in plotting a surface. Refer to Exploration Twenty-four.

PlotRange
PlotRange–>{*min*, max} specifies a range for the vertical axis in a two- dimensional plot or the third axis of a three-dimensional plot. **PlotRange**–>{{*xmin*, xmax}, {*ymin*, ymax},...} can be used to set the range on each axis of a two- or a three-dimensional plot. **PlotRange**–>**All** causes all points in the plot command to be displayed. **PlotRange**–>**Automatic** eliminates outlying points. Refer to Exploration Five.

PlotStyle
Plot[{*f(x)*, *g(x)*}, {*x*, *xmin*, *xmax*}, **PlotStyle**–>{*style₁*, *style₂*}] causes *f* to be graphed using *style₁* and *g* to be graphed using *style₂*. Style settings include **Dashing**, **Thickness**, **RGBColor**, and **GrayLevel**. Refer to Exploration Four.

Point

Point[{x, y}] is a graphics primitive that represents the point (x, y). The option setting **Epilog–>Point[{x, y}]**, for example, causes the point to be drawn in a plot. Refer to Exploration Four.

PointSize

PointSize[r] is a graphics directive which specifies that points to follow are to be shown if possible as circular regions with radius r. The radius r is given as a fraction of the total width of the graph. Refer to Exploration Four.

PolynomialQuotient

PolynomialQuotient[p, q] gives the quotient resulting from the division of polynomial p by polynomial q. Refer to Exploration Seven.

PolynomialRemainder

PolynomialRemainder[p, q] gives the remainder resulting from the division of polynomial p by polynomial q. Refer to Exploration Seven.

Prolog

Prolog is an option for graphics functions which gives a list of graphics primitives to be drawn in before the main part of the graphic is rendered. Refer to Exploration Four.

Random

Random[type, {min, max}] returns a pseudorandom number of the specified type from the indicated interval. Possible types are: **Integer**, **Real**, and **Complex**. Default range is 0 to 1. Refer to Exploration Twenty-two.

ReadList

ReadList can be used to read data into a notebook from another file. Refer to Exploration Twenty-two.

Rectangle

Rectangle[{xmin, ymin}, {xmax, ymax}] is a two-dimensional graphics primitive representing a filled rectangle with lower left-hand corner {xmin, ymin} and upper right-hand corner {xmax, ymax}. Refer to Exploration Fourteen.

Remove

Remove[symbol1, symbol2, ...] deletes the listed symbols from *Mathematica's* table of definitions. Refer to Exploration Ten.

RGBColor

RGBColor[red, green, blue] is a graphics directive which specifies that graphical objects are to be displayed, if possible, in the color or combination of colors given. Refer to Exploration Four.

Show

Show[graphics, options] displays two- and three-dimensional graphics using the options specified. Refer to Exploration Eight.

Simplify
Simplify[*expression*] transforms *expression* algebraically and returns the simplest form found. Refer to Exploration Four.

Sin
Sin[*z*] gives the sine of *z* radians. Refer to Explorations One, Twelve.

Solve
Solve[*rhs* == *lhs, x*] solves the equation for the indicated variable. **Solve**[{*equation₁*, *equation₂*, ...}, {*x₁, x₂*, ...}] solves the equations simultaneously for the indicated variables. Refer to Exploration One.

Sqrt
Sqrt[*z*] gives the square root of *z*. Refer to Exploration Nine.

Sum
Sum[*expression*, {*i, imin, imax, di*}] evaluates *expression* beginning with $i = imin$ and ending with $i = imax$ using increments of *di* and totals the results. Refer to Exploration Six.

Table
Table[*expression*, {*i, imin, imax, di*}] evaluates *expression* beginning with $i = imin$ and ending with $i = imax$ using increments of *di* and returns a table of the results. Refer to Exploration Two

TableForm
TableForm[*list*] displays *list* in a rectangular array of rows and columns. Refer to Exploration Two.

TableHeadings
An option for **TableForm**, **TableHeadings**->{{*rowlabel₁, rowlabel₂*, ...}, {*collabel₁*, *collabel₂*, ...}} places labels on the rows and columns of a table. **None** written in place of the row or the column list indicates that no labels are to be displayed. Refer to Exploration Three.

Tan
Tan[*z*] gives the tangent of *z* radians. Refer to Exploration Twelve.

Thickness
Thickness[*r*] is a graphics directive which specifies that lines are to be drawn with a thickness *r* given as a fraction of the total width of the graph. Refer to Exploration Nine.

Ticks
Ticks->{{*x₁, x₂, x₃*, ...}, {*y₁, y₂, y₃*, ...}} places tick marks on the horizontal and vertical axes of a two-dimensional plot at the positions indicated in the lists. **None** written in place of either or both of the lists will render a plot with no ticks. This option can also be applied to three-dimensional plots. Refer to Exploration Eight.

Together
Together[*expression*] puts terms in a sum over a common denominator and cancels factors in the result. Refer to Exploration One.

ToString
ToString[*expression*] prints the value of *expression*. Refer to Exploration Eleven.

ViewPoint
ViewPoint is an option for **Plot3D** which sets the point from which the surface is to be viewed. Refer to Exploration Twenty-four.

Appendix **2**

Answers to Selected Problems

Exploration 1

1. (a) −40
   (b) 1.0012
2. (a) −2.333, 4
   (b) −0.667, 1, −7
   (c) −1, 5, 2i, −2i
3. (a) $x = 0.4234, y = 0.1261$
   (b) $x = 4.65, y = 1.83; x = -4.25, y = -2.63$
4. (a)

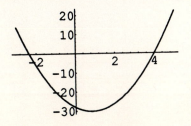

(c)

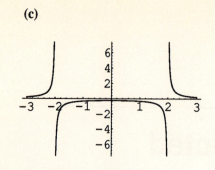

5. **(a)** $70 - 31x - 4x^2 + x^3$

   (b) $x^3 + 6x^2y + 12xy^2 + 8y^3$

   (c) $4a^7 - 20a^6b + 25a^5b^2 + 16a^6c - 90a^5bc + 125a^4b^2c + 21a^5c^2 - 150a^4bc^2 + 250a^3b^2c^2 + 5a^4c^3 - 100a^3bc^3 + 250a^2b^2c^3 - 10a^3c^4 + 125ab^2c^4 - 6a^2c^5 + 30abc^5 + 25b^2c^5 + ac^6 + 10bc^6 + c^7$

7. For example

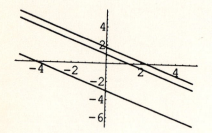

Exploration 2

1. **(a)** Max 21.3; min −344.7

3. (3.29, −354.61)

7. **(a)** 34, 950 ft^2

   (c) $30 \le x \le 266.5$

   (f) (133.25, 266.5)

Exploration 3

1. 2400 ft/sec

3. (a)

start time in min	end time in min	average velocity in units/min	end position
0	0.5	28.375	14.1875
0.5	1.	57.625	43.
1.	1.5	52.375	69.1875
1.5	2.	21.625	80.
2.	2.5	–25.625	67.1875
2.5	3.	–80.375	27.
3.	3.5	–133.625	–39.8125
3.5	4.	–176.375	–128.
4.	4.5	–199.625	227.813
4.5	5.	194.375	–325.

(c) $12t(12 - 8t + t^2)$

Exploration 4

1. The tangent slope is 10.
3. The tangent slope is 12.

Exploration 5

4. Tangent slope is –6.

Exploration 6

1. (a) Critical points occur when $x = 2.6$ and 1.4.
   (b) Critical points occur when $x = -2, -0.2$, and 1.

Exploration 7

1. There is a vertical asymptote when $x = -1$.
3. For example, $(3x^2 - 5x + 3)/(x - 1)$.

Exploration 8

1. A few features of the flight: endtime = 11.892295 min.; max. height occurs when
    $t = 7.746$ min; acceleration has a singularity when $t = 10$ min.

Exploration 9

1. (a) $c = -0.21525$
    (b) $c = 2$
    (c) $c = 0.603783$

Exploration 10

1. Yes.

Exploration 11

1. (a) Do[Plot[Sin[x] + d, {x, –2Pi, 2Pi},
                        PlotRange–>{–3.5, 3.5},
                        PlotLabel–>"d = "<>ToString[d]
                    ],
                {d, –2, 2}
                ]
3. (a) Amplitude = 3, Period = π.
    (b) Amplitude = 0.5, Period = π.

Exploration 12

1. Amplitude = 1/3, Period = 6π.
5. Amplitude = 1/2, Period = π.

Exploration 13

3. (a)

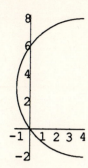

(b)

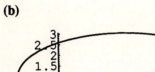

5.   maximum height = 7812.5/2 feet

Exploration 14

1.   1000 square units
5. $0.96 < A < 1.04$

Exploration 15

1.   256/3 square units
3.   3.46 square units

Exploration 16

1. (b) 67 units of distance/unit of time
3. (a) 288,000
    (b) 2400
    (c) 0

Exploration 17

1. (a) 39.84 units

    (b) 40. units

3. 1.55 units

Exloration 18

1. (a) 1/3

    (b)

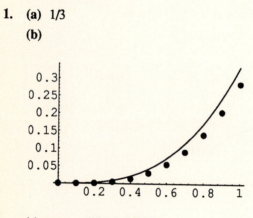

    (c) too small by 0.04833

    (d) .00498

3. 0.67

Exploration 19

1. 0.693147

Exploration 20

1.	n	LSum error	RSum error	MPSum error	trapSum error
	5.	−1.54667	1.65333	−0.02667	0.05333
	10.	0.78667	0.81333	−0.00667	0.01333
	20.	0.39667	0.40333	−0.00167	0.00333

Exploration 21

1. 2.718281828459045235
3. 2.718281785
5. 0.37 cents

Exploration 22

Half–life for the exponential curve fit to the data is 2.77 minutes.

Exploration 23

1. $y(t) = 5 - 3*e^{-t}$

Exploration 24

1. (c)

$$z = Sin[(x^2+y^2)^{(1/2)}]/(x^2+y^2)^{(1/2)}$$

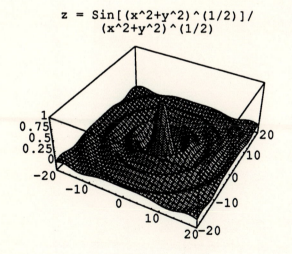

Exploration 25

1. Global maximum points at $(-1,-1,2)$ and $(1,1,2)$. Saddle point at $(0,0,0)$.
3. This surface has an infinite number of saddle points.

Exploration 26

1. Max $f = 100$, Min $f = 7.40741$
3. Max $f = 19$, Min $f = -12$

Appendix **3**

The *Exploring Calculus* Disk

The *Exploring Calculus* disk contains notebook and several files.

The Packages file contains **SecantLineAn.m**, **Area.m**, and **AreaBetweenCurves.m**, packages which are used in Explorations Five, Fourteen, and Fifteen, respectively. See those Explorations for a detailed description of these packages and the functions they contain.

The file **TrigAn** is discussed in Exploration Eleven. It contains programs which draw frames of trigonometric plots into your notebook. You can animate these frames if you wish producing moving pictures of the generationon of sine, cosine, and tangent curves.

The file called **data** contains the data points used in the study of exponential decay in Exploration Twenty-two.

Input Cells is a notebook containing every *Mathematica* input cell appearing in this book organized by exploration number. This means that you need not ever retype any of the *Mathematica* code discussed in this text if you don't wish to. All you have to do is open the **Input Cells** notebook, scroll to the appropriate exploration and double click on the cell bracket for the exploration. A cell contining that exploration's input cells will drop down the screen. You can then scroll to the particular input code that interests you, copy it and paste it into your own notebook where it can then be entered or edited just like any other *Mathematica* input cell.

Index

Learning Resources Centre